Cold War Space Sleuths

The Untold Secrets of the Soviet Space Program

Dominic Phelan (Editor)

Cold War Space Sleuths

The Untold Secrets of the Soviet Space Program

 Springer

Published in association with
Praxis Publishing
Chichester, UK

Dominic Phelan
Dublin, Ireland

SPRINGER–PRAXIS BOOKS IN SPACE EXPLORATION

ISBN 978-1-4614-3051-3   ISBN 978-1-4614-3052-0 (eBook)
DOI 10.1007/978-1-4614-3052-0
Springer New York Heidelberg Dordrecht London

Library of Congress Control Number: 2012944252

Cover design: Jim Wilkie
Project copy editor: David M. Harland
Typesetting: BookEns, Royston, Herts., UK

Printed on acid-free paper

Springer is part of Springer Science+Business Media (www.springer.com)

Contents

Foreword

By William P. Barry,
NASA Chief Historian

It is a very rare field of study that has written its own autobiography, but that is exactly what you have in your hands with this book. As a professional pursuit, the study of the Soviet/Russian space programme in the West simply didn't exist until the people who contributed to this book, along with a few others, created it. It was not easy, but these pioneers developed the necessary techniques and persevered in the face of deliberate obfuscation on one side and disheartened ignorance on the other to pursue their passion for understanding the Soviet/Russian space programme.

Space history is, of course, a new field, but after some fifty years it is coming into its own. In the early years, fans, or those with a huge enthusiasm for the technology, wrote much of what passed for space history. Respectable historians rarely touched the subject (although some, like Walter McDougall in his 1985 work *The Heavens and the Earth*, did so to tremendous effect.) In the West, outside of a few government agencies (who kept what they knew close to their collective vests), serious analysts and historians appear to have been put off by the inability to make use of the traditional tools of research when trying to decipher the Soviet space programme. Historians and participants on the Soviet side faced their own set of challenges in telling the tale. Even long after space archives had been established, they proved inaccessible, especially to Westerners. Published material was carefully edited. Memoirs, other than the largely vacuous puff pieces put out in the name of the cosmonauts, were non-existent. Interviewing participants, whose identities were often state secrets, was impossible to arrange in a systematic manner. Little wonder that public work in this field of study was left to the creativity of amateur "space sleuths" until recent years.

Each of the sleuths who contributed to this book developed skills that allowed for remarkable insights with the limited data available. With his doggedly inquisitive mind, Jim Oberg set the early standard for close reading of the published sources. Others used a personal approach to cosmonauts to methodically gather nuggets of information that helped to assemble a fuller picture. Technical skills, like the ability to find and track spacecraft radio signals, were used by some of the sleuths. Others honed their Russian language skills and pored through so much material that they could find the inconsistencies and telling errors in the carefully managed stream of propaganda. To me, one of the most remarkable skill-development stories here is that of Phil Clark – a man who largely taught himself the math and computer skills needed to do sophisticated and stunningly accurate analyses of orbital mechanics. Whatever talent they brought to bear, each of these sleuths showed remarkable tenacity in pursuing a better understanding the Soviet/Russian space programme.

While the individual accomplishments in these stories are remarkable, what is most striking is the degree to which an international community of interest grew out of these efforts. There was certainly a fair amount of competition among the sleuths, and a few old grudges are evident in these pages. However, long before the Internet-age made it easy, this group of enthusiasts from some seven different countries (and others beside them) discovered one another. Initially gravitating to the remarkable Geoffrey Perry and his Kettering Group, the Soviet space sleuths eventually found an institutional home of sorts in the British Interplanetary Society. In the pages of its *Spaceflight* magazine and at the annual Soviet Forum, the sleuths soon found that the sum of their efforts was much greater than the parts. How do you explain the creation of such a fertile cooperative effort among such disparate and admittedly competitive researchers? It was not merely a willingness to collaborate for the greater good, nor the outlet for their work provided by the society. To my mind a critical role was played by the selfless mentorship of early leaders like Geoff Perry and, later, Rex Hall. When I had the opportunity to indulge my passion for Soviet space sleuthing in the early 1990s, Nick Johnson introduced me to Rex. From the warm reception and encouragement that I received, you would have thought I was Rex's long lost brother. But, Rex and his ever-patient wife Lynn, seemed to treat everyone that way. Without the help of Rex, Phil Clark, and the rest of the Soviet space sleuths, I certainly would never have been able to make my small contribution to the literature. The value of selfless mentorship and commitment to a community of effort is a powerful lesson that you will find reinforced in the stories collected in this book.

One of the other delightful aspects of this book is that the stories complement one another, but still present a unique perspective. Like the apocryphal story of the blind men and the elephant, each of the contributors tells the story from their point of view. The result is that the enigma of Soviet/Russian space history is conveyed from a number of angles. Such variations add to the richness of our understanding of space history, and also to our understanding of the founders of the Western study of the Soviet/Russian space programme. Having known most of these authors for the last twenty years, I continually found myself surprised (and a bit embarrassed) to learn new details about them and their struggles to pursue their passion.

One theme repeated throughout these essays is the question of whether the age of "space sleuthing" is over. There is no doubt that the situation has changed radically since the collapse of the Soviet Union over twenty years ago. It is now possible for both Russians and those of us in other parts of the world to use some of the more traditional tools of historic research and analysis. While the (partial) openness of Russian archives, and the availability of participants and their papers, is largely a result of political change, our knowledge of the questions to ask and the ability to put the answers into perspective is largely a result of the efforts of the Soviet space sleuths. Those that contributed to this book, and the many that passed away before their stories could be told, created the field of Western Soviet/Russian space history. Yet, it is also clear that their work is not done. While the field may now be more mainstream, and allow for the use of more traditional analytic tools, it still poses a number of unique challenges. Creativity, tenacity and many of the tools developed by the early space sleuths will still be required when dealing with conflicting stories and limited access to issues that were highly classified state secrets. There is plenty of sleuthing left to do. Those of you just getting started, as well as many of us who have been around awhile, could do nothing better than to read this autobiography of the field. Happy hunting!

Editor's introduction

The idea for this book came to me only days after the death of Rex Hall – the legendary host of the British Interplanetary Society's annual 'Soviet Forum'. Over the years he had become the chief moderator for a tight-knit international network of amateur space historians called the 'space sleuths'. Although this group has lost five well-known researchers recently, to me his passing really did seem to symbolise the end of an era.

Within weeks I was in contact with many of his friends and colleagues in search of their 'war stories' about the golden era of space sleuthing – many of which I include in my chapter. For me this task was an honour, as my own fascination with their efforts dated back to October 1986 when, as a fourteen-year-old astronomy student, I first spotted the now iconic *National Geographic* with a saluting, spacewalking cosmonaut Leonid Kizim on the cover. Ironically, 1986 ought to have been memorable for a host of exciting space events, including the launch of a new Soviet space station, a flotilla of space probes to Halley's Comet and the launch of the Hubble Space Telescope, but instead it is now remembered as the year of the Challenger disaster. As a profoundly shocked West looked on, an invigorated USSR under the dynamic leadership of Mikhail Gorbachev seemed (finally) to be winning the 'Space Race' – an impression enhanced by the launch of their giant Energiya super booster in May 1987.

What really caught my eye about that launch was the fact that the above mentioned *National Geographic* article contained a detailed drawing of the top-secret booster six months before it was revealed to the world. Quickly rereading the article introduced me to now familiar names such as Geoffrey Perry, Rex Hall, James Oberg and Charles Vick. Luckily, my growing interest in their work was aided by the chance discovery of a complete set of the British Interplanetary Society's *Spaceflight* magazine from the 1970s in a second-hand bookshop in Dublin – allowing me to cherry-pick a decade's worth of the best articles from the 'golden era' of space sleuthing. In those days (before the Internet), it really did feel like you were an amateur spy if you tuned into Radio Moscow to catch the latest space news. My notebooks from that period also show my interest in how this news was reported in the West, as they mention various appearances by space sleuths in the media

0.1: *Cold War Space Sleuths'* editor Dominic Phelan.

explaining what it all meant. This interest would later influence my decision to study Journalism in college – something that I hope gives this book a slightly different angle on the usual story of Soviet spaceflight.

During 1988 the publication of three books on the subject – Phillip Clark's *Soviet Manned Space Programme*, James Oberg's *Uncovering Soviet Disasters*, and Brian Harvey's *Race into Space* – provided references to new books and articles for me to track down. The discovery that Brian Harvey also lived in Dublin and gave regular talks to the Irish Astronomical Society there now seems serendipitous. Thankfully he turned out not to be over-protective of his knowledge and became a mentor, encouraging me to attend the BIS' Soviet Forum in London. Although I didn't attend until 1993, it was still possible to meet legendary figures such as Rex Hall and Phil Clark in person, and whilst some of the sleuths' politics might have ranged from ideological 'Fellow Travellers' to 'Cold Warriors', most of them were simply in it to apply their detective skills to discovering the truth behind the public face of the Soviet space programme. You might ask yourself what makes me uniquely qualified to be the editor of this book, when I haven't revealed any previously unknown Soviet space secrets myself? Coming from a younger generation that missed most of the actual 'sleuthing' gives me, I believe, the slight detachment needed to judge their work objectively.

Although space sleuthing itself might now seem like 'history' too, working on *Cold War Space Sleuths* has given me renewed grounds for optimism that the craft is alive and well. Currently, the most promising future target is undoubtedly China. This rising superpower is not only using Soviet-style technology, but in many ways is also copying its media control techniques. Also, Russia might revert to its centuries-old tradition of attempting to hide embarrassing facts from foreigners when co-operation aboard the ISS comes to an end. Apart from these future possibilities, those lucky enough to be searching through the Russian archives tell me that there are plenty of undiscovered documents waiting to be found which could potentially rewrite our understanding of some aspects of spaceflight history.

The space sleuthing discussed in this volume might be the product of a certain age, but the skills they pioneered might come in useful again during the 21st Century!

Dominic Phelan,
Dublin, Ireland

Acknowledgements

Naturally, it has been a pleasure working on this book because I never imagined that one day I would get the opportunity to ask many of my writing heroes to contribute a chapter to a book with my name on the cover. *Cold War Space Sleuths* is the type of book I've always wanted to read myself, and I'm just lucky that I got there first with the idea to ask the sleuths to put their stories on paper.

My sincere thanks then to: Asif, Bart, Bert, Brian, Christian, Claude, Dave, Jim, Phil and Sven for their enthusiastic responses when I told them about the proposal to produce this book.

The list of those who didn't contribute a chapter but whose input was just as valuable is as important, and I would also like to acknowledge the help of: NASA historian William Barry for the Foreword; Peter Smolders, Lynn Hall, Robert Christy, Leo Enright, Marcia Smith and Michael Cassutt for background information; Charles P. Vick for permission to use his famed Cold War-era rocket drawings; although the photographs used in each chapter are the copyright of the individual authors, I would also like to thank Martin Dawson, Alistair Scott, Gerald Borrowman, Charles Vick, Bert Vis and Ed Cameron for some additional hard-to-find images that were used; Ralph Gibson/RIA Novosti for the photograph on the front cover; Stephen Corbett and Ed Zigoy for providing vital *Spaceflight* and *JBIS* back issues; Colin Burgess (my own editor on *Footprints in the Dust*) for making me realise that anything is possible if you can assemble the right team; Mary Todd and Suszann Parry of the British Interplanetary Society for their tireless efforts to ensure the Soviet/Chinese Forum runs smoothly each year; former *Spaceflight* editor Clive Simpson for publishing some of my earlier articles; and Clive, Maury, Romy, Jim and David at Springer-Praxis for making the publication process run smoothly.

Finally and most importantly a big thank-you to all my family, especially my parents Beryl and Tony, my brother Damien and sister Lisa, for their continued support and interest in my writing. Enjoy the book!

1

Space sleuths and their 'scoops'

by Dominic Phelan

The USSR was famously described by Winston Churchill as "a riddle, wrapped in a mystery, inside an enigma" and nothing signified this more than the search for the truth behind its space programme during the Cold War. Although the Space Race was literally played out above our heads, it was often obscured by a figurative 'space curtain' that took much effort to see through. Although Western governments with the latest spy technology at their disposal often had a pretty good idea of the real story they tended to keep their findings to themselves. As a result, to find out what was really going on, amateurs often had to rely on their own skills at reading between the lines of official Soviet announcements.

BEHIND THE IRON CURTAIN

The first time Westerners noticed the Soviet's seriousness about spaceflight was a 1951 newspaper article by scientist Mikhail Tikhonravov [1]. Around that time the first Western speculation on Soviet rocket technology also emerged in the form of an influential paper presented to the American Rocket Society [2]. At the Californian 'think tank' RAND, analyst Firmin Krieger had started a detailed survey of Soviet newspaper articles seeking definitive proof of serious Kremlin interest in spaceflight [3], but it was Russian-born Colgate University academic Albert Parry who predicted that the first satellite would probably be launched close to September 1957 in order to mark the centenary of space theorist Konstantin Tsiolkovsky's birth [4].

The first English-language book devoted to the subject of Soviet space technology was published in 1961, but unfortunately Alfred Zaehringer's slim volume is now only notable as an example of how little Western writers had to work with in trying to paint a picture of what was going on behind the Iron Curtain in the early days of the Space Race [5].

The situation had somewhat improved by the time prolific American aerospace writer Martin Caidin penned *Red Star in Space* [6]. Although technical details were

1.1: Soviet secrecy only provoked sceptics like author Lloyd Mallan.

still difficult to find, Caidin rightly acknowledged that the Americans had nobody but themselves to blame for ignoring Soviet intentions in space. At the other end of the spectrum was ultra-sceptical American journalist Lloyd Mallan, who firmly believed that the Soviets were lying about everything [7]. He didn't believe they had launched anyone into space!

'Space Sleuths' – early media creation

The first time the unofficial Soviet space watchers were termed "space sleuths" was in a 1963 *Popular Mechanics* article [8]. Although it noted that there were about a dozen part-time space analysts in the US, it focused on Donald J. Ritchie of the Bendix Corporation and his efforts to reconstruct the still-secret Vostok spacecraft and its launch vehicle.

Unfortunately these speculations were far from reality and his drawings included such imaginary features as a large tailfin with four rocket cylinders attached, a sealed internal cockpit for the pilot, and solid-propellant retrorockets. "I think people did their best to try to get what they could," believes space author Brian Harvey, whose own interest was sparked as an eight-year-old boy when he saw Vostok 2 fly over Ireland in August 1961. "At the Red Square parade there would be a helicopter flying past with a model of Vostok underneath it – with a big ringed tail. These models would appear in the Western press as what Vostok looked like, but the accompanying commentaries were generally responsible, as they would say that this is best that we knew. The people who did know what it looked like were the CIA but they weren't going to tell."

All these early non-covert speculations suffered from the fact that the only way to try and reconstruct the Soviet launch vehicle was to observe the left-over fuel tanks placed into orbit with the satellite. This mainly involved guessing its dimensions by the brightness as it streaked overhead. Thankfully, the third stage of the vehicle was already known because it was often exhibited in the West attached to models of the early Luna probes. These fragmented clues, combined with the long-held view that the Soviets were using upgraded German V2 rocket technology, gave the impression of a booster that was very different from the reality. It was only when the Soviets revealed the true shape of their workhorse R-7 rocket at the Paris Air Show in 1967 – after it was militarily obsolete – that the Western sleuths realised their mistake. They soon discovered that Russian approaches to technology often differed markedly from those imagined.

A Cold War bluff?

In Britain, television astronomer Patrick Moore and writer Kenneth Gatland were the first to make important contacts with Soviet scientists at various international conferences. Both men had connections to the famed British Interplanetary Society (BIS), with Moore editing its magazine *Spaceflight* before Gatland took over the job in 1959. The society's reputation for exposing Soviet space secrets was made in 1963 when Russian defector Grigory Tokaty-Tokaev used one of its lectures to reveal that

1.2: (L-R) Defector Leonid Vladimirov pictured with space sleuths Phil Clark, James Oberg and Rex Hall in 1988.

the previously unknown Sergei Korolev was an important figure. Britain at that time was an important link between the two space powers, with the giant Jodrell Bank radio-telescope just outside Manchester becoming the focus of some early scoops, mainly because the Soviets had the habit of informing its director, Bernard Lovell, whenever they launched a new lunar probe!

Although a few good books on Soviet spaceflight were available around the time of the tenth anniversary of the launch of Sputnik [9], the appearance of two 'insider accounts' packed with new information in the early 1970s caused a sensation. After his defection to Britain in 1966, Russian science journalist Leonid Vladimirov tried to publish a book revealing that the Soviets were behind in the Space Race but he soon found that publishers just didn't believe him and didn't want to risk printing a book that would be out-of-date if the Soviets won the Moon Race as then imagined. It was only after the triumph of Apollo 11 that he was finally able to find a market for his work, with *The Russian Space Bluff* appearing in 1971. Strangely, many respectable Western experts refused to take the book seriously [10].

To give it a balanced review, *Spaceflight* magazine's new editor Kenneth Gatland turned to Rolls Royce engineer Arthur 'Val' Cleaver but unfortunately he, too, was so taken with the long list of Soviet space spectaculars that he failed to see the truth in some of the defector's claims [11]. His scepticism motivated professional historian

Robert Conquest (a long-standing BIS Fellow) to write a letter in defence of the defector's view of the USSR [12].

A more significant book was *The Kremlin and the Cosmos* by American correspondent Nicholas Daniloff [13]. "Daniloff was a native Russian speaker and an accomplished journalist who did an extensive amount of research on the topic," notes space historian Asif Siddiqi. "Despite some missteps, about the Moon Race, for example, his book still stands up 40 years later as a remarkably good introduction to what was known about the Soviet space programme in the early 1970s."

Asif Siddiqi, a Bangladeshi-American now regarded as one of the pre-eminent historians of the subject, believes several distinct groups of space sleuths existed. "My feeling is that there have been three generations of researchers. The first from the 1950s and 1960s included Krieger, Sheldon, Perry, Shelton, Stine, Daniloff, and Zaehringer. The second from the 1970s and 1980s included Clark, Gibbons, Grahn, Hall, Hooper, Kidger, Lardier, Smolders, Covault, Johnson, Oberg, Vick, Wachtel, Woods, and Zaloga. The third from the 1990s included Haeseler, Harvey, Hendrickx, Przybilski, Shayler, Villain, Vis, Wotzlaw, Barry, Gorin, Harford, Newkirk, Pesavento, Podvig, Wade, and Zak."

These sleuths can also be classified into three distinct groups: those writing in the English language (a diverse group that included not only British and American but Irish, Scandinavian, Belgian and Dutch sleuths); continental Europeans writing exclusively in French and German; and a small band of native Russian historians whose research efforts would only start to emerge in the 1990s after the demise of the USSR. The three groups worked in isolation owing to language barriers – often resulting in them independently coming to the same conclusions.

THE KETTERING GROUP

In the early 1960s a whole division of amateur radio enthusiasts sprung up around the globe tuning into 'secret' Russian satellite transmissions. Ironically their interest was sparked by the early Soviet decision to choose radio frequencies for the Sputniks that could easily be picked up by amateurs as proof that they were successful. One of the first to master the techniques needed to track Soviet satellites was Kettering Grammar School teacher Geoffrey Perry. He originally started tracking Sputniks as a means of inspiring his science class, but soon became so adept at it that he often scooped the professionals at tipping off the press to the latest Soviet mission. A grateful media soon termed his team the "Kettering Group". Over the years, this informal group would grow to include well-known satellite trackers such as Robert Christy, Max White and Fritz Muse in England, Sven Grahn in Sweden, and Chris van den Berg of The Netherlands.

Listening to satellite signals

By the mid-1960s Perry had noticed that some satellites were being inserted into different orbits, and inferred that a new launch site was being used. By "walking

1.3: Geoffrey Perry tracked satellites from his classroom.

back" their orbital tracks to a starting point on the map, he was able to discover its location near Archangel in northern Russia. Although the Plesetsk Cosmodrome was well-known to the CIA – the U2 flown by Gary Powers had been shot down trying to photograph it – Perry was the first in the West to tip-off the media about it when he mentioned it at a BIS lecture in November 1966 and wrote a follow-up letter to *Flight* magazine. Shortly thereafter, *Time* magazine reported his "discovery" [14].

Strangely, he wasn't the first to use this method to discover a launch site. Back in 1957 an astronomer at Tokyo Observatory had used the first two Sputnik flights to pinpoint their launch site in Soviet Central Asia – when many still thought they were being launched from inside Russia. Unfortunately his scoop wasn't noticed because he only published it in Japanese [15].

A friend in Washington DC

Someone who benefited greatly from the work of the Kettering Group was Dr Charles S. Sheldon II. His own fascination with spaceflight started when he met German rocket pioneer Wernher von Braun at White Sands, New Mexico, in the 1940s. By the 1960s Sheldon was chief of the Science Policy Research Division of the Congressional Research Service (CRS) and was tasked with preparing objective, non-partisan reports on the Soviet space programme for the US Congress.

These volumes – published from the 1960s to the 1980s under the general title *Soviet Space Programs* – were the most informative works on the subject in the

1.4: Charles Sheldon with young protégés James Oberg and Charles Vick in 1974.

English language during the Cold War. A major handicap for Sheldon was the fact that he could only use material already printed in the open literature, so he did his best to encourage amateur sleuths in their own researches [16].

"Sheldon had full security access, but could only use non-classified material for his congressional reports," remembers American sleuth James Oberg. "Without ever once transgressing his security boundaries, he encouraged our investigations and speculations in directions along which he knew in advance we'd find pay dirt. When he published our BIS papers in his own reports, we felt verified – and we had been."

Although Sheldon died in September 1981, his reports continued, thanks largely to the on-going detective work of the sleuths he had encouraged.

"The last two of the series, 1976-1980 and 1981-1987, were written after Charles died and I can state authoritatively that they could not have been completed without Geoff and the Kettering Group," notes Sheldon's former research assistant Marcia S. Smith.

REPORTER'S EYE FOR DETAIL

Over the years, The Netherlands has produced more than its fair share of good space sleuths. The first was Peter Smolders, a journalist fluent in Russian as a result of his training with Dutch Army Intelligence during his national service from 1961 to 1963.

1.5: Dutch journalist Peter Smolders, with back to camera, interviews cosmonaut boss General Kamanin in 1970.

With this skill, Smolders was able to tease facts from his hosts during dozens of trips to the Soviet Union researching space stories.

"As a Dutchman I was somewhere in between the US and Russia, both geographically and culturally," believes Smolders. "The fact that I was well received had, I think, to do with my objectivity – especially in the sixties and seventies when there was a lot of unpleasant propaganda and politics involved from both sides. During the Cold War I did not feel like a spy but more like a detective. The Dutch intelligence service wanted me to work for them. I refused but, at the same time, I felt lucky to live in a normal country where I was able to do that."

Soviet newsreels provide clues

Slowly 'secrets' began to slip through to the West because of the increasing number of official Soviet newsreels then being released. They often contained interesting clues for those willing to watch them frame-by-frame. Luckily for Smolders, his connections with the authorities gave him access to 35-mm Sovexportfilm newsreels. "I had a simple hand-operated device with film reels, and could check every individual frame. I looked at a scene where Soyuz cosmonauts were walking in the RKK Energiya museum (then still closed to the public) with the Voskhod 2 capsule in the background. And there, out of focus, on the wall I discovered a scheme of the Voskhod spacecraft – and that became the basis of my drawing in *Soviets in Space*."

His book was first published in The Netherlands in 1971, with the English translation appearing in 1973 [17]. The main revelation of his drawing was the location of an extra retrorocket on the top of the Voskhod capsule – itself merely a modified Vostok. During the IAF congress held in The Hague in 1974, Smolders was surprised when Alexei Leonov drew a sketch of Voskhod 2 with the extra retrorocket and called it "Eta Voskhod Smoldersa" [18]. "So he had seen the drawing that we did for the English-language book published in 1973. Up to that moment, Leonov had only made fantasy drawings of the spacecraft, as he was not allowed to show the real thing. Later of course he did his realistic oil paintings of his spacewalk."

Cosmonaut Musa Manarov would later tell Smolders that they had seen a copy of his book in the library of RKK Energiya, and were always amazed at the amount of information known in the West.

Proton rocket in the frame

The release of the Soviet movie *Steep Road to Space* (1972) about the triumph and tragedy of Salyut 1, provided Smolders with the first hint of the layout of the then secret Proton booster: "I analysed a film of the launch of the Proton carrying the first Salyut. I could see only the upper part of the rocket, but there was one frame (just one!) showing the tips of the boosters. Had I just looked at the films on the screen, I would never have discovered those details. It must have come as a surprise to the Russians!"

But even this first glimpse was deceptive, because what in reality was a novel arrangement of six cylindrical fuel tanks wrapped around a central core tank was wrongly interpreted.

Smolders' drawing of the booster (which was first published in a Dutch magazine when he was in Florida covering the Apollo 17 mission in December 1972) shows the booster as a giant version of the tapered Vostok rocket but with six strap-on boosters instead of four. A long-range view of the rocket also shown in *Steep Road to Space* confused matters further by appearing to show that the central 'core stage' wasn't firing at lift-off.

It took the skills of experienced nuclear industry draftsman Charles Patrick Vick to create a more accurate representation of the Proton. His 1973 drawing had cylindrical strap-on boosters, and this interpretation became the accepted configuration until the truth was revealed in 1984 [19]. At that time it was believed in the West that this rocket started out as a giant intercontinental ballistic missile (ICBM) developed by missile designer M. K. Yangel [20]. Westerners never guessed that what appeared to be strap-on boosters were really external fuel tanks surrounding a central core oxidiser tank; reminiscent of the design of NASA's first giant rocket, the Saturn I. The Proton (UR-500) had been conceived as a giant missile to carry a huge 30 megaton nuclear warhead and its novel arrangement of propellant tanks was dictated by the size limits imposed on what could be carried around the country on the Soviet railway network.

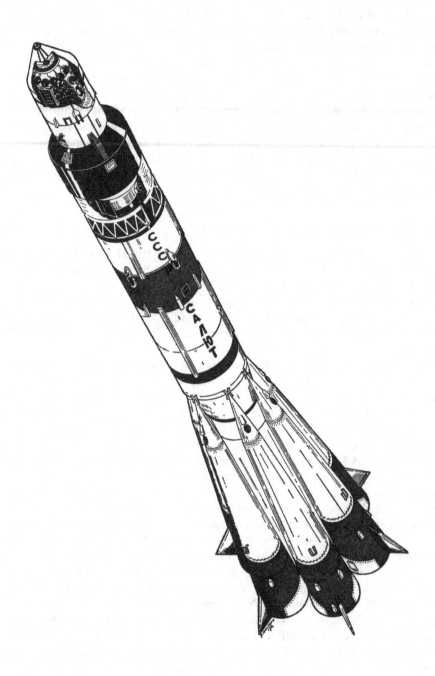

1.6A/B: The first Western drawings of the Proton were published by Peter Smolders and Charles Vick.

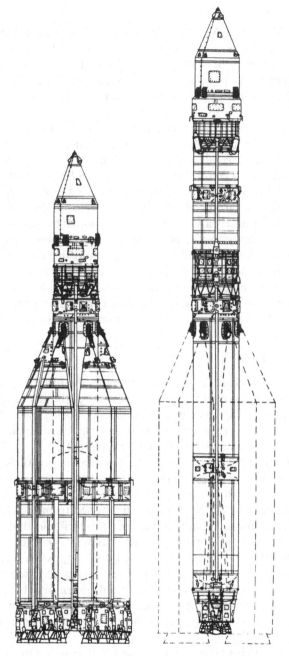

The third man

Peter Smolders contacts in Moscow also provided him with a further grim scoop when he was told the truth about the deaths of the Soyuz 11 cosmonauts. During an interview with cosmonaut Boris Yegorov in September 1971 he was informed that the crew died when a pressure valve accidently opened above the Earth's atmosphere. NASA wouldn't be told this until training had begun for the Apollo-Soyuz Test Project (ASTP) some time later.

During this period, rumours also began circulating that Leonov and Kubasov had originally been the prime Soyuz 11 crew, together with an unknown third cosmonaut. Owing to the last-minute replacement of this prime crew by their backups, it was an open secret in Moscow that Alexei Leonov had escaped death. It didn't take long for this to leak to Westerners visiting the Russian capital to report on the ASTP mission. Bizarrely, proof existed on the front cover of the January 1975 issue of *Spaceflight* for nearly a year before it was spotted by sharp-eyed readers. Just before the mission, the Soviet censor released an old photograph of what they thought was the Leonov-Kubasov crew without noticing the presence of another man's forehead in the lower right-hand corner [21]. When Western space sleuths realised that this was the original Soyuz 11 crew they wrote a stream of letters to *Spaceflight* trying to figure out who it was; but their guesses were wrong, and it would turn out to be a previously unknown cosmonaut named Pyotr Kolodin.

A game of 'Space Cluedo'

In the early 1970s American sleuth James Oberg realised that embarrassing facts in Soviet newsreels had been overlooked by the censors. Although he ended up working at NASA in the mid-1970s, Oberg was still with the US Air Force when he started to search through early newsreels seeking 'missing cosmonauts' amongst the familiar faces [22]. This work would quickly gain Oberg a reputation as the *Enfant Terrible* of space sleuths for a series of articles ridiculing Soviet attempts to hide their existence.

Several months after the publication of the first article on the subject, his attention was drawn to two versions of the same photograph appearing to show a cosmonaut whom he labelled 'X2', crudely airbrushed from the cosmonauts' ranks [23]. Oberg correctly guessed that the man was 'Grigori' and that he had probably been dismissed because of serious character flaws [24]. Not long afterwards, *Spaceflight* reader Rex Hall in London mentioned a possible cosmonaut named 'Nelyubov' [25]. It wouldn't be until 1986 that the Soviets admitted that a cosmonaut named Grigori Nelyubov had been fired after a drunken brawl.

But Oberg's writing style had its detractors – with one contributor to *Spaceflight* complaining that Oberg was playing a "politically motivated game of Space Cluedo" [26].

Irish space author Brian Harvey believes part of the fun of Oberg's work is that he likes "the game". As Harvey puts it, "He is provocative, interesting, adversarial and controversial but he does have a basic respect for the facts."

MOON RACE MYSTERY

James Oberg was one of the main advocates of the then-controversial view that the Soviets had been in the race to the Moon, and used his new position as an associate editor of the US magazine *Space World* to pen several lengthy articles on the subject. These were later refined into the essay 'Russia Meant to Win the Moon Race', which went on to win the National Space Club's prestigious Robert Goddard Space History Award – even though NASA historian Eugene Emme recommended that he place a question mark at the end of the title [27]. By that time Western analysts had already guessed the existence of two separate parallel lunar projects: a circumlunar manned Zond launched on the existing Proton booster and a lunar landing mission that would require a secret giant heavy-lift rocket [28].

During this period Oberg also became NASA's "unofficial ambassador to the Russian space programme" whenever a cosmonaut paid a visited to Houston. It was often left to Oberg to show the visitors around the Johnson Space Center because his bosses felt uncomfortable doing so [29].

Starting in the late 1970s, Oberg also contributed high-profile columns for *Omni*, the science magazine published by *Penthouse*-founder Bob Guccione, which brought space sleuthing to a wider readership. In fact, his profile became so big that *The Right Stuff* author Tom Wolfe provided a foreword to his first book on Soviet spaceflight.

The "missing giants"

The existence of a giant rocket to match America's Saturn V would have been indisputable evidence that the Soviets had been in the Moon Race. Although such a rocket was well-known to the American intelligence community, the development of the "N-1 booster" would not be admitted by the Soviets until 1989. Rumours of its spectacular launch failures appeared in *Aviation Week & Space Technology* (in the process confirming its nickname of 'Aviation Leak'), and prompted Charles Sheldon to refer to them mischievously as the "missing giants" [30].

Despite the fact that solid proof of the giant Soviet booster in the open literature was scarce, most Western sleuths were sufficiently informed to know that it had to exist. Although the 20th anniversary of the Apollo 11 landing is often regarded as the main spur for an increasingly open Soviet society to reveal their past lunar ambitions, it is no coincidence that the controlling influence of Valentin Glushko had just ended with his death in January 1989. Others could finally challenge his "official history" of the late 1960s – something he had easily managed by serving as editor of the official *Soviet Encyclopedia of Cosmonautics* [31].

At first, many were still reticent to reveal details of a still-sensitive blank page in space history. But in the same month that Glushko died following a prolonged illness, Vasily Lazarev became the first cosmonaut to publicly admit to a lecture audience in Perm that the lunar programme existed. Even so, he was reluctant to go into details, saying that it was "a matter of the past" [32].

The first solid facts came in a book about cosmonaut Valery Bykovsky published that summer, which told the story of his training for a manned Zond flight in the 1960s. In December 1989 a group of visiting professors from the Massachusetts Institute of Technology (MIT) were finally shown a real Soviet lunar lander during a tour of the museum of the Moscow Aviation Institute. On their return to the US they passed their pictures to the *New York Times*, which published them on the front page under the headline "Russians Finally Admit They Lost Race to Moon".

Accurate drawings?

When the first tentative proof of the Soviet manned lunar project emerged in 1989, Russian newspaper weren't able to illustrate their articles with official photographs of the N-1, as these were still classified! Ironically, they used Charles Vick's drawings. Although Vick was not the first Westerner to publish a depiction of the booster that bore a satisfactory resemblance to its true shape – G. Harry Stine illustrated one of his articles with a large rocket consisting of three distinctively conical stages, but we will probably never know whether this was a lucky guess on Stine's part or the result of a tip-off [33] – his obsession was finally paying off.

Vick had been attempting to reconstruct the secret Soviet booster since the early 1970s. Then he was honest enough to admit that there simply wasn't enough evidence in the open literature for him to draw an accurate representation of what he called the "Lenin booster" but what was also designated the "G" booster under a now-obsolete Sheldon classification system [34]. Luckily by the end of that decade Vick managed to discover an important clue to the giant Soviet booster's true shape when he spotted a drawing of an unknown launch tower that was inexplicably published in a Soviet book entitled *Kosmodrom*. We now know this book had been edited by some Korolev loyalists, so in hindsight, perhaps they were trying to slip an acknowledgement of its existence under the nose of the Soviet censor. Whatever the reason for its publication, the drawing caught Vick's trained eye and he created a rocket whose shape would fit the contours of the tower.

"(The book) is dated 1977 and I got it in the fall of that year or early 1978," Vick remembers. "Sheldon was stunned by the declassification and had a lot to say about it to me. At the time I was working for Bechtel nuclear power plant and saw a lot of oil rig piping structures, so I took that gantry and started on reverse engineering it. The service levels indicated the stage areas diameters or inter-tank areas, but what was bogus was the suggested length of the first stage, which I now know harkened back to the N-1 nuclear rocket configuration described in the early 1960s."

Unfortunately for Vick, he was not destined to be the first to reveal the booster's true shape to the public, because *Aviation Week* reporter Craig Covault included an illustration of an expected future Soviet heavy-lift booster in an article in June 1980. Obviously based on a high-level US intelligence briefing, we can now recognise this illustration as being the N-1 design [35]. Despite that, when Charles Vick's new N-1 drawing appeared in Kenneth Gatland's *The Illustrated Encyclopedia of Space Technology* in 1981 it caused a sensation at the Moscow Book Fair. When an official Russian translation was published later, Vick's embarrassing drawings were edited out [36].

Vick went on to explain his reconstruction of the Soviet lunar rocket design at the 1983 BIS Soviet Forum [37] but many fellow sleuths were sceptical that he could have come up with the conical shape without Sheldon's help. Vick is adamant about his scoop: "[To] suggest that I changed the N-1/Type-G concept based on *AW&ST* is absolutely incorrect. We did it as two separate efforts based on our own sources and methods."

With the existence of the giant Soviet booster now acknowledged by Western researchers, attention turned to finding a matching Soviet lunar lander. Observers suspected three obscure Cosmos flights of having tested such a vehicle. D. R. Woods and Sven Grahn used the orbital movements of these craft to imagine what the lander might look like [38]. Belgian researcher Theo Pirard spotted what he thought was the secret spacecraft in a hazy launch pad photograph of a 1969 Soyuz mission, but it wasn't the secret lunar lander being shown to Soviet pressmen as hoped. Writing in *Spaceflight*, London-based émigré engineer Martin Postranecky reported that he had seen this object on the cover of a 1971 Czech-language magazine and it was merely a futuristic commentary box built for Russian television [39].

Despite the overwhelming evidence, some Western space analysts only accepted the existence of a Soviet manned lunar programme after it was officially admitted by Moscow. "I had my 'Russian scoops', but I failed to estimate the amount of effort devoted to the Russian manned moonlanding effort until *glasnost* and my meeting with former chief designer Vasily Mishin," admits Peter Smolders.

LONDON RENDEZVOUS

Since *Spaceflight* magazine had been used by the sleuths of the 1970s to discuss the latest ideas about Soviet space, it was fitting that the British Interplanetary Society's headquarters in Vauxhall, South London, would become the unofficial meeting place for the sleuths in the 1980s.

British researcher Anthony Kenden was one of the first to promote the study of the Soviet space programme as a serious academic exercise, and in order to assist others with their research he published a guide to the printed material available [40]. It was mainly through his efforts that the Society organised its first all-day 'Soviet Forum' in January 1980, with the papers being published in a special 'Soviet Astronautics' issue of the *Journal of the British Interplanetary Society (JBIS)*. Although Rex Hall later became the main host of the event, Kenden played an early role. Unfortunately, owing to his death aged 40 in a car accident in 1987 his contribution has been largely forgotten.

Rising stars

Another notable Soviet Forum regular was Phillip Clark. Back in 1969, as he watched the Apollo 11 mission on television, Clark wonder why the "so-called experts" had gotten it so wrong predicting the outcome of the Moon Race [41]. To find out more he contacted Kettering Group founder Geoffrey Perry, and by 1976

1.7A/B: The first officially released N-1 photograph bares a strong similarity to this Charles Vick drawing from the 1980s.

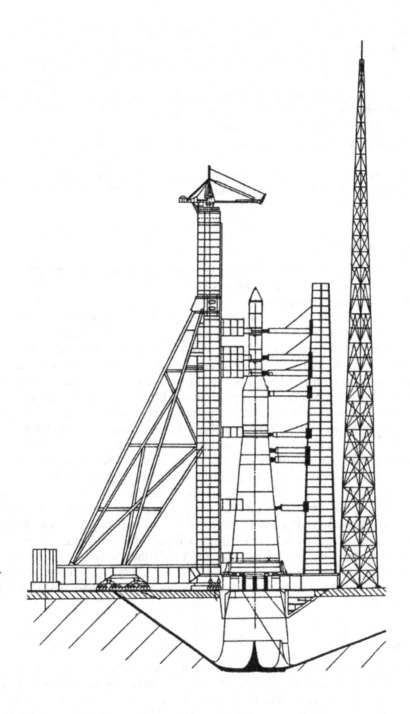

was applying his mathematical skills to accurately predict the durations of manned missions to Soviet space stations before they were even launched [42].

The publication of James Oberg's book *Red Star in Orbit* in 1981 can be seen as an important event in the history of space sleuthing because it served to inspire a new generation of researchers with its gripping chapters on the deadly 'Nedelin' rocket explosion in 1960, the secret Moon Race, and on-going work to identify the missing cosmonauts [43]. As Brian Harvey recalls, "His book is a terrific read and I read it twice through in six hours! Somebody had actually put the whole thing down in writing and persuaded a publisher to take it – that was the significant bit."

A decade later Oberg's work was being recognised in Russia itself and he had the thrill of seeing *Red Star in Orbit* on prominent display in a Moscow space museum [44].

The French connection

During the Cold War, France kept itself between the two competing superpowers to such an extent that many important figures in the Soviet space programme felt safe attending the famous Paris Air Show.

Christian Lardier and Claude Wachtel were part of a research group centred on the 'Cosmos Club de France' (C2F) and pioneered the systematic study of Russian-language material in the search for clues overlooked by others. Ironically, their own findings were often overlooked by other sleuths because they were published only in the French-language *Orbite* journal. For this reason Claude Wachtel caused a minor sensation in 1981 when he presented a paper at the BIS Forum revealing previously unknown Soviet spacecraft designs he had found in the Russian book *The Creative Legacy of Sergei Korolev* (1980). This was the first time any of the English-language sleuths had seen this material, and it altered their understanding of the early history of the Soyuz project. It also provided some of the first concrete evidence that the Soviets had thought seriously about sending cosmonauts to the Moon [45].

The French sleuths also pioneered the search for the real identities of the "Chief Designers". Using the simple rationale that most of them were Academy of Sciences members and Communist Party officials, they started a detailed search through party decrees and obituaries. It was then just a painstaking process of matching their real names with the fake official pseudonyms which they were obliged to use in public. As a result, many of the entries on Soviet Chief Designers in *Cosmos Encyclopédie* edited by Albert Ducrocq were more accurate than their English-language rivals.

Real spying?

Before the Internet era, the BIS Forum helped to keep all the international researchers in direct contact with each other. "Although it was the *British* Interplanetary Society, there were times when the number of people coming from outside of Britain would be greater," notes Brian Harvey. He attended his first meeting in 1982 and was there to hear a telegram from James Oberg reporting that

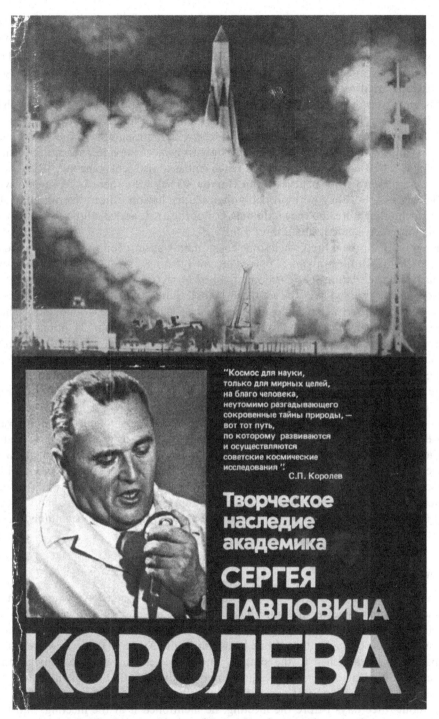

1.8: *The Creative Legacy of Sergei Korolev* caused a sensation.

Cosmos 1374, a 'mini-shuttle' test, had just splashed down in the Indian Ocean. Returning home, Harvey mentioned this to Irish television journalist (and BIS Fellow) Leo Enright – only to be asked "What Russian shuttle test last week?" The story was confirmed the following year when the Australian Air Force released photographs that it had taken of a second recovery in March 1983 (Cosmos 1445). "This improved my credibility as you can imagine! Leo put me on television that evening, saying 'Dubliner finds Russian space shuttle'," laughs Harvey.

Bizarrely, those early BIS meetings also saw a representative from the Soviet embassy in the audience – officially there only to project the latest space film which the embassy had agreed to show. The French embassy also sent someone. "Rex knew they were there to spy," believes Brian Harvey. "They were there as representatives of the embassy but they were probably the military liaison officer. We weren't told exactly who they were and they didn't have a lot to say. It added a little *frisson* to the whole thing and made it much more fun!"

Eventually this backfired when some cosmonaut transmissions picked up by radio tracker John Branegan were played to a meeting. His tapes included the embarrassing revelation that cosmonauts liked to record Country & Western radio stations as they passed over the southern United States and then transmit the songs back down to their tracking ships off the coast of Cuba and Newfoundland. "They were extraordinary tapes, and were fabulous stuff, but within 24 hours of our meeting the transmissions had gone coded," remembers Harvey. "The explanation is the fairly obvious one that the nice embassy official who had brought us the films had gone back to the embassy, rang Moscow, and asked them if they were aware that there were all these guys in London who had tapes of our cosmonauts chattering and would they please introduce some radio discipline!"

One of the acknowledged masters of listening in to the cosmonauts was Chris van den Berg, who learned the art of clandestine radio monitoring as a teenager listening to the BBC from Nazi-occupied Holland. He could even tell when a cosmonaut was spacewalking by the 'goldfish bowl' sound of their voice. "He was very successful in the second half of the 1990s when he was following the Mir downlink," recalls fellow Dutch sleuth Bert Vis. "He discovered several things that the Russians hadn't even told NASA yet, and that they only found out when reporters were beginning to ask questions based on his MIRNEWS newsletters!"

A new openness

Thankfully traditional old-style Soviet secrecy began to break down when Mikhail Gorbachev came to power and introduced the policy of openness known as *glasnost*. "In 1985, about three days before the launch of Soyuz-T 13, when we suspected that Salyut 7 was in serious trouble, I had heard on Radio Moscow that a new crew was preparing to be launched to the station. And they were named as Savinyikh and Dzhanibekov," Harvey recalls. "Radio Moscow didn't say that this is the crew that is going to fly, but if you put two and two together you did get four. So I rang up Rex about three days before the meeting, he mentioned it at the meeting, and the launch took place with that crew about two days after. I think that was the first ever

1.9: BIS Soviet Forum 2008 speakers: (L-R) Bert Vis, Bart Hendrickx, Gerald Borrowman, Andrew Ball, Dominic Phelan and Brian Harvey.

time that a crew had been known in advance. If you had asked anyone sitting there in the 1980s if they would believe that in a decade someone would be there telling about their visit to Baykonur they would be sending for the people in white coats."

Small successes like this encouraged Harvey to think about writing his own book about the Soviet space programme, so he hired an electronic typewriter and used the large collection of press clippings that he had amassed since the early 1960s [46]. "I think I got it largely right but I'm not claiming the credit for that [because] I listened very carefully to the analysis that I was getting at the BIS and tried to reach my own view," says Harvey. "I would attribute a lot of my knowledge and my judgement of what went into *Race into Space* from having been to the BIS meetings. I think it gave me a lot of confidence and it certainly gave me both breadth and depth that I could not have gotten anywhere else."

FRONT COVER TREATMENT

National Geographic first came up with the idea of a major 'Soviets in Space' feature at the time of the 20th anniversary of Yuri Gagarin's historic flight in 1981. Science writer Thomas Canby was sent to Moscow in June 1982 to exploit the rare welcome extended to the Western press during a joint Franco-Soviet manned spaceflight, but unfortunately his article was shelved after the shooting down of the South Korean airliner KAL 007 in September 1983 [47]. During this period Canby's enthusiasm for the project was kept alive only by "gentle nagging" from space sleuth Charles Vick, who often appeared with a large folder of drawings under his arm. Fortunately for both, Canby was able to persuade the magazine's editorial board to resume work on the article in late 1985.

A year of tragedy and triumphs

Ultimately, fate ensured that the finished *National Geographic* story gained a much wider readership than if it had been published as originally scheduled. By the time that Canby returned to Moscow to finish his research in March 1986, the world had changed in many respects. The West was still in shock at the recent explosion of the space shuttle Challenger, which resulted in an almost existential crisis of confidence in NASA, whilst the Soviet Union, under its dynamic leader Mikhail Gorbachev, had launched its latest space station Mir (Peace). The mainstream press might have been surprised by the new station's Gorbachev-friendly name rather than the dull 'Salyut-8' they expected, but it was no surprise to sleuth Peter Smolders, who had been told the new name during a trip to Moscow a year earlier [48].

   National Geographic's renewed approaches to the Soviet press agency Novosti were initially rebuffed, as it had opened doors four years previously and had then been embarrassed when the magazine did not publish the promised article. But after some discussion Novosti assisted the American writer, providing an interview with Vladimir Dzhanibekov, recently returned from rescuing Salyut 7. As Canby displayed mockups of how the magazine feature might look, the cosmonaut kept a diplomatic silence about its numerous illustrations of still-secret rockets and space shuttles.

The space shuttle race

The *National Geographic* story finally appeared as a cover story in late 1986 [49]. The West was suddenly a passive observer to a string of Soyuz liftoffs shown on live television, and speculation of the imminent introduction of a Soviet space shuttle now seemed credible. That year, Nicholas L. Johnson's annual *Soviet Year in Space* noted that the "disparity" between the two superpowers was at its greatest since 1957, with the USSR accounting for a staggering 88 per cent of the total launches in 1986. It is not surprising that over the next few years the mainstream US media published high-profile stories on the subject [50].

   Whilst all these articles credited the Soviets as being ahead in long-term manned spaceflight experience, access to heavy-lift rockets and the number of launches per year, they all reminded their readers that the Soviets had nothing which matched the sophistication of the American space shuttle. Since the inaugural launch of Columbia in 1981 the West had been awaiting what the press dubbed 'shuttleski', although the sleuths had known since 1983 that it was to be called 'Buran' (Snowstorm) [51].

There was almost a media frenzy in the run-up to its launch, because it coincided with the return to flight of the grounded US shuttle fleet after the loss of Challenger. Although some commentators thought it might have been launched during President Ronald Reagan's Moscow visit in May 1988, most presumed it would be launched purely to steal some thunder from Discovery's imminent flight. However, we now know that Buran's designers were desperately trying to launch it before the project was cancelled by an increasingly sceptical Kremlin. In an attempt to boost Buran's profile, they promised live television coverage of the launch but technical problems caused delays. Their original launch plans, which would have seen it take off into a clear blue daytime sky were thwarted by technical problems and it eventually lifted off in darkness. As it quickly disappeared into the low clouds, the Soviet shuttle was denied a spectacular photo-opportunity for the cover of *Pravda*, and official interest evaporated even quicker. Although it grabbed some headlines when it was flown on the back of a carrier aircraft to the 1989 Paris Air Show, this 'other shuttle' was never launched again [52].

Sharing information

Privately circulated newsletters also played an important role in enabling sleuths to exchange information before the Internet era. Perhaps the most interesting of these was the Dutch-language *Spaceview* magazine produced in the mid-1970s. Although only a two-man operation by Maarten Houtman and Jacob Terwey, a member of the Netherlands-Soviet Union Friendship Society, it soon gained a worldwide readership because of it numerous scoops. These were mainly the result of their willingness to travel behind the Iron Curtain to attend communist-sponsored youth events in search of contacts. During these trips, they often met cosmonauts and space journalist who had just returned from Star City or Baykonur and gave first-hand accounts of what they'd just seen. Houtman and Terwey also had contacts at the Novosti press agency, which gave them the latest pictures. This resulted in their spotting interesting things that slipped by the Soviet censor. Their most famous find was a picture which proved the fatal Soyuz 1 flight had been planned as a two-ship docking mission. [53].

Unfortunately for the magazine's long-term future, Maarten Houtman was also an active member of the Dutch squatters' movement and in 1978, during a raid on the squat in which he lived, the magazine's entire archive was tossed onto the street and destroyed by the riot police. It was relaunched in the 1980s under the editorship of Luc van den Abeelen, but never had the same impact.

The 1989 BIS Soviet Forum saw one of its largest ever attendances, with over fifty people heeding Rex Hall's call not only for more openness from the Soviets but also

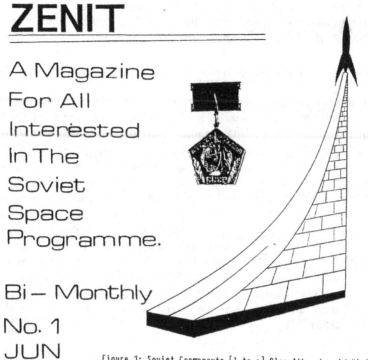

ZENIT

A Magazine
For All
Interested
In The
Soviet
Space
Programme.

Bi – Monthly

No. 1

JUN
1985

Figure 1: Soviet Cosmonauts [l to r] Oleg Atkov, Leonid Kizim
and Vladimir Solovyov speak with Georgi Beregovoi shortly
after their record breaking 237-day spaceflight which ended
on 2 October 1984.   An article by Dr. Atkov appears in
this issue of ZENIT: Photo courtesy "Zemlya i Vsellenaya".

1.10: The first issue of the journal *Zenit*.

for Western researchers to share information with their colleagues in the new era of "semi-*glasnost*" [54].

One of the best outlets during this period was the A5-sized fanzine-type journal *Zenit* which ran from 1985-1991, published in England by David J. Shayler under the editorship of Neville Kidger. It only had some 200 subscribers at its peak, but it was enormously influential amongst sleuths by making it possible to rapidly disseminate the latest revelations coming out of Moscow. "Although *Zenit* didn't last very long, it was a wonderful resource," recalls Asif Siddiqi. "I remember reading every issue and [was] just stunned with the amount of information coming out." Another important reference for any self-respecting space sleuth was Nicholas Johnson's annual *Soviet Year in Space*.

Nigel McKnight's picture-based *Spaceflight News*, which was also published from 1985-1991, was an excellent source of contemporary Soviet space news. 'SFN' as it was known to its readers, also led to the British Interplanetary Society redesigning its *Spaceflight* magazine and launching it on the newsstands in early 1987.

Flight International was always a good source of material, and by the late 1980s its 'spaceflight page' compiled by Tim Furniss was providing an important outlet for many of the space sleuths to air their latest speculations. Furniss took part in one of the first 'press junkets' to Baykonur. In 1966 General de Gaulle was the first official Westerner to tour the launch site, and in that tradition François Mitterrand was invited to watch the lift off of the latest French cosmonaut on a Soyuz mission in 1998. What added spice to the trip for the press delegation was that it took place only ten days after the launch of Buran. They had hoped to get to see the Soviet shuttle during the trip, but that privilege was only extended to the VIPs. An attempt by one French journalist to gate crash the trip to the shuttle was frustrated only because his bus got lost in fog [55].

By then *glasnost* was giving Western analysts the answers to questions they had been asking for years, but it was also signalling the end of the 'golden era' of space sleuthing. By the late 1980s the Mir space station had almost become a household name in the West, whilst a symbolic parity had been reached with the launch of the Buran shuttle and a quote from a 'Soviet space expert' in the media was common. James Oberg was the main expert for the American media, whereas in Britain both Phil Clark and Geoffrey Perry were often quoted. Clark was becoming a regular on the BBC, which sent him to Moscow in 1988 to report on the forthcoming Fobos missions. This high-profile was maintained by the publication of his first book in time for Christmas 1988 [56]. Clark later acted as a technical consultant for a three-part documentary called *Red Star in Orbit* broadcast in December 1990.

On his retirement from Kettering Grammar School in 1984 Geoffrey Perry was hired by ITN as its official 'space consultant', and he would even gain a measure of mainstream recognition in 1987 when Channel 4 filmed his story for a documentary drama called 'Sputniks, Bleeps and Mr. Perry'. After the publication of his book *Race into Space* in 1988, Irish author Brian Harvey appeared regularly on state broadcaster RTE as a commentator on the latest space news.

By then *glasnost* had its own downside. Recriminations over the loss of the two Fobos probes, criticism of the Buran shuttle inside Russia, and the revelations about the 'Nedelin explosion' in 1960 were quickly taking the gloss off Soviet spaceflight's stellar reputation. And the near tragic ending of a joint Soviet-Afghan propaganda flight didn't help matters [57].

The official archives open

As a young space fan, Bart Hendrickx's attentions had turned to the Soviets during the long lull in the American space programme caused by delays in developing the US space shuttle.

"I started following the missions to Salyut 6, mainly by listening to the daily reports on the cosmonauts' activities on Radio Moscow World Service. Sometimes I would hear excerpts of air-to-ground conversations and was frustrated that I didn't understand a word of what was being said. Having had a lifelong interest in foreign languages, the next natural step was to take up Russian," remembers Hendrickx. "I didn't get to go to Moscow all that often in the late 1980s [and] early 1990s but whenever I did go there I did, of course, scour the bookshops for anything of interest. Actually, most of the big revelations in the early days were not made in books but in regular newspapers like *Pravda* and *Izvestia*, and magazines available on newsstands here in Belgium and also in a Russian bookshop in Brussels that I regularly used to go to. The problem has been that the really interesting books are published only in very small [numbers]. Some of the design bureau histories seem to be written for employees of the design bureaus themselves rather than for outsiders. You won't even find them in bookshops in Moscow and you often need to find contacts to get hold of them. But that's part of the fun, looking for and finding the books and then realising that you're one of only a handful of people in the West to have them."

A series of articles written by Hendrickx translating and interpreting the diaries of the head of cosmonaut training in the 1960s are still considered his most important contribution to the field [58].

"It was pretty labour-intensive but probably the most fun I've ever had doing this kind of work. Amongst the articles I've written, these were the ones that sparked the most responses, so in the end it was very rewarding," reveals Hendrickx.

Another Western historian who has specialised in using Russian-language material is Bangladeshi-American Asif Siddiqi. In 1995 he talked NASA's History Office into a definitive book on Soviet spaceflight based on Russian primary sources, and then he taught himself the language over the next several years. *Challenge to Apollo* received wide acclaim inside Russia [59]. Later Siddiqi was asked to edit the diaries of Boris Chertok, one of the central figures of the early Soviet space programme.

"The information published in the history of the Soviet space programme in the Russian language is vast, just enormous, and we have only begun to touch the tip of it in the English-speaking world," explains Siddiqi. "And then if you start to go deeper, into the actual archives, it's another ocean of materials. I've had the great fortune of working in some of these archives over the past few years and I can tell you that the information they have is just amazing. I have seen with my own eyes, for example, hundreds of pages of documentation on the abandoned UR-700 manned lunar project which have never been published, even in Russian. There are still things waiting to be discovered."

Siddiqi acknowledges the assistance of a new generation of Russian researchers that included the late Vadim Molchanov, Sergey Voyevodin, Timofey Varfolomeyev, and *Novosti kosmonavtiki* writers Igor Lissov, Igor Afanasyev, Sergei Shamsutdinov, Igor Marinin, Vladimir Agapov and Konstantin Lantratov. Other Western researchers working with Russian material include US-based analyst Pavel Podvig (Soviet ICBM development); German researcher Dietrich Haeseler (Soviet rocket engines), Olaf Przybiliski (N-1 and the Almaz military space station), and Matthias Uhl (V2 rockets in the Soviet Union after World War II).

Definitive histories?

Despite the opening up of the Russian archives, some in the West retain a degree of scepticism. American research Peter Pesavento, one of the first to use the Freedom of Information Act to obtain access to previously classified US intelligence documents, is of the opinion that *these* sometimes can offer a better picture of Soviet intentions during the Cold War [60]. "Some [people] have apparently seized to their chests the notion that Russian releases since 1990 are of Western-quality standard," Pesavento says. "US intelligence documents now being released after being in the processing mode for years, show that what the Russians have told [us] about their space history is very incomplete in [many] aspects." Perhaps the best example of his scepticism is the two-part re-examination of the history of the Moon Race that he co-authored with Charles Vick [61].

As Bart Hendrickx admits, "There's so much fascinating material waiting to be discovered in the archives, but design bureau archives are largely off-limits even to Russian researchers, not to mention Western ones. With some effort it is possible to gain access to some of the state archives but, again, you must have the time and the resources to do this. Unfortunately being a space historian is not a paid job! And let's face it: I guess there are not all that many people with a real passion for the intricacies of Soviet space history, so you have to be willing to spend long hours working on articles that will ultimately be read by a relatively small audience."

By 2000, most Western space historians went directly to Moscow to get answers to any final questions they had about the early days of spaceflight. Dutch researcher Bert Vis was one of the first to make contacts inside the cosmonaut training centre. He started visiting the once-closed facility in 1992. He arrived at just the right time, and was able to interview many obscure unflown cosmonauts before they died. Rex Hall often accompanied Vis on these research trips and the two even had the honour of being invited to place a wreath at Yuri Gagarin's grave in the Kremlin Wall on 'Cosmonautics Day' in April 2001. Thankfully, Hall had the chance to co-author *The First Soviet Cosmonaut Team* with Australian writer Colin Burgess before his death from cancer in 2010 [62]. "At least once a week I would receive a large envelope stuffed with rare photographs and obscure information that [went on to become the] heart of this book," recalls Burgess. "Rex always suggested that he didn't contribute much to this book, but he was wrong. Although he is no longer with us to refute my feelings on the subject, he poured his experience, personal insights, wisdom, personal contacts and a beloved, lifelong collection of Soviet-Russian material into this book."

CONCLUSIONS

Perhaps the last word should be given to James Oberg, the man who more than most became the archetypal Cold War Space Sleuth.

"I'm astonished we got so much right, and I'm delighted that we missed so much that our post-Soviet Russian colleagues still had a lot to teach us," says Oberg. "I still get chills remembering how I lived long enough, and the USSR died soon enough, for me to stand on the scoured concrete apron in Area 41 where Nedelin's team perished, or to hold in my hand a fragment of the July 1969 N-1 failure, or to shake the hand of a cosmonaut [who had been] erased from early photographs. Awesome beyond belief. Of all the futures anticipated by me as a young man, none of these events were within range of even the wildest hopes."

References

1. Brian Harvey, 'Mikhail Tikhonravov: His Contribution to the Soviet lunar and Interplanetary programme', *JBIS* Vol. 59, p. 266.
2. George P. Sutton, 'Rockets Behind the Iron Curtain', *ARS Journal*, May-June 1953, p. 189.
3. Firmin Krieger, *A Casebook on Soviet Astronautics (2.Vol)*, RAND Corporation 1956. Re-issued *Behind the Sputniks*, Public Affairs Press, Washington DC 1958.
4. 'Albert Parry, Russia Expert, 92; Predicted Sputnik 1 Launching', *New York Times*, 8 May 1992. James Oberg, *Red Star in Orbit*, Harrap, London 1981, p. 40.
5. Alfred Zaehringer, *Soviet Space Technology*, Harper Bros, New York 1961.
6. Martin Caidin, *Red Star in Space*, Crowell-Collier Press 1963.
7. Lloyd Mallan, *Russia's Space Hoax*, Science & Mechanics Publishing 1966.
8. 'Space Detectives Rate Russia's Moon Probes', *Popular Mechanics*, March 1963, p. 89.
9. William Shelton, *Soviet Space Exploration – The First Decade*, Washington Square Press 1968.
10. Leonid Vladimirov, *The Russian Space Bluff*, Tom Stacey London 1971.
11. A. V. Cleaver, 'The Russian Space Bluff', *Spaceflight*, June 1972, p. 202.
12. Robert Conquest, 'The Russian Space Bluff', *Spaceflight*, November 1972, p. 437.
13. Nicholas Daniloff, *The Kremlin and the Cosmos*, Alfred Knopf, New York 1973.
14. G. E. Perry, 'The Soviet Northern Cosmodrome', *Spaceflight*, August 1967, p. 290.

15. Charles S. Sheldon, 'The Soviet Space Program Revisited', *TRW Space Log* 1974, p. 3]
16. 'Personal Profile: Dr. Charles S. Sheldon II', *Spaceflight*, August 1971, p. 290. G. Perry, 'Charles Stuart Sheldon II, An Appreciation', *JBIS* Vol. 35 No. 2, p. 50.
17. Peter Smolders, *Soviets in Space*, Lutterworth, London 1973.
18. Peter Smolders, 'Those Missing Cosmonauts', *Spaceflight*, May 1975, p. 172.
19. William Broad, 'Space Sleuth keeps eye on Soviets', *New York Times*, December 25 1984.
20. C. P. Vick, 'The Soviet Superbooster – 1', *Spaceflight*, December 1973, p. 457. James Oberg, *Flight International*, 16 August 1973, p.323. Phillip Clark, 'The Proton Launch Vehicle', *Spaceflight*, September 1977, p. 330.
21. James Oberg, 'Was Soyuz 11 Flown by the Back-up Crew?', *Spaceflight*, June 1976, p. 236.
22. James Oberg, 'Missing Cosmonauts from the Class of 1960', *Spaceflight*, August 1973, p. 309.
23. James Oberg, 'Soviet Trainee Cosmonauts', *Spaceflight*, February 1974, p. 42.
24. James Oberg, 'Russia's Manned Space Programme for the 1970s', *Spaceflight*, October 1974, p.398.
25. Rex Hall, 'Missing Cosmonauts', *Spaceflight*, February 1975, p. 79.
26. Geoff Clayton, 'Politics and Space Cluedo', *Spaceflight*, January 1976, p. 38.
27. James Oberg, 'The Moon Race was Real', *Space World* March 1975 p. 4. 'Alexei Leonov', *Space World* June 1975 p.10. 'Russia Meant to Win the Moon Race', *Spaceflight*, May 1975, p.163.
28. Nicholas L. Johnson, 'Apollo and Zond – Race around the Moon, *Spaceflight*, December 1978, p. 403.
29. James Oberg, *Star-Crossed Orbits*, McGraw-Hill 2002, p. 34.
30. Charles S. Sheldon, 'The Soviet Space Program Revisited', *TRW Space Log* 1974, p. 11.
31. Dominic Phelan, 'The Eagle and the Bear', *Footprints in the Dust, p.* 75, University of Nebraska Press 2010.
32. 'B. Povarnitsin' letter, *Spaceflight*, November 1989, p. 390.
33. G. Harry Stine, 'Some strange things happened at Baykonyr', *Analog*, October 1970, p.107.
34. C. P. Vick, 'The Soviet Super Booster – 2', *Spaceflight*, March 1974, p. 94. James Oberg, 'The next man on the Moon', *Analog* February 1975, p. 73.
35. Craig Covault, 'Soviets Developing 12-Man Space Station', *Aviation Week & Space Technology*, 16 June 1980, p. 26.
36. Kenneth Gatland, *The Illustrated Encyclopedia of Space Technology*, Salamador Books London 1981, p. 40 & 53. Phillip Clark, 'The Soviet N1 Manned Lunar Programme', *Zenit* (No. 34) December 1989, p. 5.
37. Charles P. Vick, "The Soviet G-1-e Manned Lunar Landing Programme Booster," *JBIS*, Vol. 38 No. 1, p.11, 1985.
38. D. R. Woods, 'A Review of the Soviet Lunar Exploration Programme', *Spaceflight*, July-Aug 1976, 287. Woods, 'Lunar Mission Cosmos Satellites',

Spaceflight, November 1977, p. 383. Sven Grahn, 'Lunar Module Tests', *Spaceflight*, October 1973, p. 398.

39. 'Soviet Lunar Module?', *Spaceflight*, November 1976, p. 398. 'A Mystery Solved?', *Spaceflight*, March 1977, p.118.
40. Anthony Kenden, 'A Guide to the study of the Soviet space programme', *Spaceflight*, May 1975, p.175.
41. Phillip Clark, 'The Russian Space Riddle', *The Observer*, 4 December 1988 p. 29.
42. Phillip Clark, 'Soyuz Missions to Salyut Stations', *Spaceflight*, June 1979, p 259.
43. James Oberg, *Red Star in Orbit*, Harrap, London 1981.
44. James Oberg, 'Soviet Space Secrets', *Spaceflight*, August 1995, p. 254.
45. Claude Wachtel, 'Design Studies of the Vostok-J and Soyuz Spacecraft', *JBIS* Vol. 35 No. 2, p.92.
46. Brian Harvey, *Race into Space*, Ellis Horwood, Chichester 1988.
47. Thomas Canby, *From Botswana to the Bering Sea*, Island Press 1998, p.179.
48. James Oberg's *Cosmogram* newletter, No.15, 23 November 1985.
49. Thomas Canby, 'Soviets in Space: Are they Ahead?', *National Geographic*, October 1986.
50. 'America Grounded', *Newsweek* 17 August 1987. 'Surging Ahead: The Soviets overtake the U.S. as the No.1 Spacefaring Nation', *Time* 5 October 1987. 'Red Star Rising', *U.S. News & World Report*, 16 May 1988.
51. *Cosmogram*, No.6, 20 December 1983.
52. Bart Hendrickx & Bert Vis, *Energiya-Buran*, Springer-Praxis 2007.
53. James Oberg, 'Soyuz 1 Ten Years After: New Conclusions', *Spaceflight*, May 1977, p.187.
54. 'Society Symposium Success', *Spaceflight*, October 1989, p.349.
55. Tim Furniss, 'Behind the Curtain', *SpaceFlight News*, February 1989, p.29.
56. Phillip Clark, *The Soviet Manned Space Programme*, Salamandor Books, London 1988.
59. 'Soyuz TM5: what went wrong', *Flight International*, 15 October 1988, p.42.
58. Bart Hendrickx, 'The Kamanin Diaries 1960-63', *JBIS* Vol. 50 No. 7 p.33. 'Diaries 1964-1966', *JBIS* Vol. 51 No. 11 p.413. 'Diaries 1967-168', *JBIS* Vol. 53 No. 11/12 p.384. 'Diaries 1969-1971', *JBIS* Vol. 55 No. 9/10 p.312.
59. Asif Siddiqi, *Sputnik and the Soviet Space Challenge & The Soviet Space Race with Apollo*, University of Florida Press 2003.
60. Peter Pesavento, 'Soviet Space Programme: CIA Documents reveal new historical information, *Spaceflight*, July 1993, p. 224.
61. Peter Pesavento & Charles Vick, 'The Moon Race End Game', *Quest* Vol. 11 No. 1 (Part 1), Vol. 11, No.2 (Part 2), 2004.
62. Rex Hall and Colin Burgess, *The First Cosmonaut Team*, Praxis Publishing 2009.

2

Hidden in plain view

by Brian Harvey

It was a clear warm late summer's evening when I first saw it. A bright star appeared from the west and steadily, silently crossed the night sky heading from the Wexford coast in south-eastern Ireland, bending over the sea toward England and Europe. 'There's a man up there!' I was told (only the second person to fly into orbit) but this brief encounter with Vostok 2 and Gherman Titov was full of meaning. The Russian space programme, whose defining public characteristic was its secrecy, was not just something one followed in the newspapers: you could actually see it happening.

FIRST LOOK BEHIND THE SPACE CURTAIN

That summer, two publishers rushed to get into print accounts of the events of that spring. Panther published, in Wilfred Burchett and Anthony Purdy's *Gagarin*, not only an account of Yuri Gagarin's mission, but first-hand research and interviews done in the Soviet Union. They wrote of people we had not heard of much before, scientists like Iosif Shklovsky and Mstislav Keldysh, of lunar rovers, Venus probes and canine missions. They explained the romantic sense of mission that had driven the Russian space programme from the 19th century. All this was a revelation. The second publication was Pyramid's *Man into Space*, by Martin Caidin, a prolific, patriotic American writer on the wartime air force and its postwar rocketry. His account captured the atmosphere of early-sixties Cape Canaveral like few others, and it still makes enthralling reading today. His heroic account of the flight of Alan Shepard concluded with two short but provocative chapters on the Soviet programme, one an admiring commentary on Yuri Gagarin's flight and the other on Sputnik. He chronicled all the many Soviet announcements from 1951 to September 1957 leading up to Sputnik, taking several pages to do so, before asking: "So, just WHAT was so secret about the first Russian satellite?"

2.1: Brian Harvey with cosmonaut Georgi Grechko in 1991.

"Thick-fingered peasants"

The accounts by Burchett and Purdy and by Caidin undermined the received wisdom of the day, which was that the Soviet space programme was so hopelessly secret that one could never gain more than a superficial picture of its history, present activities or future intentions. They showed what, with an intelligent bit of probing, could be done. Of course, the programme was secret in so many ways. Launchings were not announced until they took place. Cosmonaut names were not released until they flew. Mission details were sparse. We still did not know the size or shape of their rockets or cabins or most of their spaceships, except for the first Sputniks, lunar probes and Venus probe. This was fertile ground for rumour, speculation and worse. *Reader's Digest* published the story of two radio hams in Turin, Italy, who claimed to have picked up signals from dying cosmonauts in spaceships hurtling out of control ('Amateurs with their eyes on space', May 1965).

The Guinness Book of Records, which also ought to have known better, wiped out another squad of cosmonauts in missions from 1957-1961; their long legacy was such that even in 2011, the 50th anniversary of Gagarin's flight, people were still asking about 'missing' cosmonauts. All these events took place against the background of the Cold War, which had a distorting effect on the narrative of the early Soviet space programme. Even granted its less than open nature, their programme hardly deserved the unflattering terms in which it was then reported.

Their rockets were crude. They had stolen designs from Germany and the Americans. Cosmonauts were expendable. They never published their scientific results. As Caidin pointed out, the Russians were seen as "thick fingered peasants with terrible electronics fumbling their way around into space". And so on. Indeed, these epithets resurface nowadays – in reporting of the Chinese space programme. The first challenge for the amateur, as Caidin had illustrated, was to get behind the conventional narrative and dig deeper.

A Soviet challenge?

All this created a challenging field worthy of investigation. The first problem that faced any investigator in this early period was to make an accurate assessment of Soviet intentions in space, especially in response to the American Apollo project. The media always wanted to know: *What are the Russians up to now? What are they going to do next?* The chief interlocutor was Bernard Lovell, director of Jodrell Bank, whose giant radio-telescope could pick up signals from so far out in space that even the Russians travelled to Manchester in the spring of 1961 to try to find their missing first Venus probe. In the autumn of 1963 Lovell became not merely an interpreter but an actor in this Cold War drama, for in reciprocation for his tracking efforts he was invited to visit the Soviet Union. There was consternation when, counter-narrative, he returned with the opinion that there was no active plan to race the Americans to the Moon – correctly, as the government decision to do this was not made until August 1964. This challenged the widely held belief that the Soviet Union must be striving to be first to reach the Moon as part of its programme of global communist conquest. This same narrative also envisaged an efficient command-and-control economy of successive five-year plans in which Kremlin leaders clicked their fingers and design institutes whirred into action and spaceships duly took off.

Only occasionally did the mainstream Western media make serious attempts to interpret the Soviet space programme. Kenneth Gatland of the British Interplanetary Society (BIS), who edited its magazine *Spaceflight,* was often published in the *Sunday Telegraph.* The *Evening Herald* ran 'Mystery of Soviet space key figure', trying to figure out who really led the Soviet space programme. It concluded that this was Valentin Glushko, which was a good try (he was the next one). Such behind-the-scenes analyses though, were rarely sought by the mainstream media.

CIA leaks

A second challenge was to be alert to the occasions when the wall of secrecy might break down, as walls of secrecy tend to do. In August 1964 a French newspaper ran a story that the USSR would shortly launch three cosmonauts into orbit together. The story gained little traction, probably because such a venture was considered to be well beyond the capabilities of the time, but the mission flew two months later as the first in the Voskhod series.

A new player appeared on the scene when the Central Intelligence Agency (CIA) published a record of failed Soviet launch attempts to the Moon, Mars and Venus, so numerous as to cast doubt on its credibility (in the event, the list proved to be entirely accurate). The CIA reports were rarely published in the agency's name, but the details were provided informally to American newspapers and periodicals, principally *Aviation Week* (jokingly called 'Aviation Leak' as a result). Over the years, CIA commentaries and assessments, or those parts that were released, proved to be extraordinarily accurate, but in the hot-as-a-greenhouse Cold War narrative of the time it was not always easy to decide who to believe. The American magazine *Newsweek* used CIA sources regularly, making it an early, reliable source of interpretation of the Soviet space programme. *Newsweek* was uncannily accurate in predicting the capacity of the Proton launch vehicle to send a manned spaceship on a loop around the Moon and back to Earth – exactly the profile of the subsequent Zond mission ('Biggest boost', 2 August 1965). The failure to rival Apollo 8 in December 1968 was due to "unspecified problems with Zond", which was an honest admission in the absence of more specific information, but also fundamentally accurate ('Soviet moon shot postponed', 16 December 1968).

BREZHNEV'S THAW

In the event, the change of government in October 1964 brought a thaw, a dropping in the temperature of the 'hot' Cold War and a slightly more open approach by the Soviet Union. Although the reforming government of Leonid Brezhnev is not well regarded now, it did bring greater openness as well as collaboration not only with the socialist bloc but further afield, especially France – an initiative which could only lead to an improved flow of information. In 1965 the Vostok cabin was displayed for the first time. In 1966 the identity of the chief designer, Sergei Korolev, was at last revealed following his death. In 1967 the Vostok launch vehicle was revealed at the Paris Air Show. That same year, the American television company NBC was allowed to film cosmonauts in training and a commemorative film was issued on the first ten years of the space programme.

Ultimately, it was important for an investigator to try come to closer quarters with the programme itself. Most Western reporters never got any closer than the well-choreographed post-mission press conferences. Here, Soviet publications became important. The best points of access were left-wing bookshops in Britain and Ireland, and the Soviet embassy in London. The main publications were the monthly *Sputnik* (broadly comparable to the *Reader's Digest*) and *Soviet Weekly*. *Sputnik* carried the first detailed accounts of the life of Sergei Korolev, as well as detailed accounts of individual missions (e.g. Luna 16 with the first surface sample and Luna 17 with the first Lunokhod) with quality photographs.

Having to sell on Western news shelves (even if they were in leftist bookshops) required higher attention to production values, especially in the quality of paper and illustrations. 'Tracks on the moon' (*Sputnik*, April 1972) reported a day-visit to the Lunokhod control centre and quoted a selenologist called Alexander Basilevsky, an unknown name then but now the leading selector of landing sites for missions to the Moon and planets. The magazine included human interest stories such as 'Valentina

2.2: The *Sputnik* cover story about Sergei Korolev.

Gagarina tells of Yuri's last day' (*Sputnik*, January 1969) and many years later 'The girls who didn't go' – the story of the first squad of women cosmonauts.

Reading *Soviet Weekly*

Soviet Weekly provided a layer of detail not available in the mainstream Western publications, especially on the interplanetary missions, and often featured diagrams, cutaway drawings and illustrations. *Soviet Weekly* was the first, for example, to publish a detailed diagrammatic exposition of the Mars 2 and 3 missions ('How the soft-landing on Mars was accomplished', 15 January 1972). It also published some off-beat articles, such as Valentin Lebedev's lengthy account of life in orbit ('Diary of a cosmonaut', 10 September 1983).

The embassy was also the outlet for occasional publications such as *Soviet News* and *Soviet Booklets* which might be themed to report particular missions. Later, the range of published material in the bookshops widened, one of the most useful being an annual *Soviet Science & Technology Almanac* which reviewed space missions of the previous year. 1969 saw publication of Пересадка на орбите (*Transfer in orbit*), a lavishly illustrated book on the Soyuz 4 and Soyuz 5 linkup, providing a level of detail never seen before.

In 1973 the Academy of Sciences published Valentin Glushko's *Development of Rocketry and Space Technology in the USSR*, a short but definitive history of rocket engines, with much new detail on scientific missions. It also included illustrations of some of the key personalities of the programme (e.g. Mikhail Tikhonravov) and even prominent engineers who were shot by Stalin (e.g. Ivan Kleimenov) – albeit saying little about them or the circumstances of their demise (the sharp observer, however, would observe that many died in 1937 or 1938). Additional details were given in his sequel, *Rocket engines GDL-OKB* (1975).

The *Soviet Booklet* series could be divided into those that were nicely illustrated but largely hagiographical; and those that were genuinely informative and provided fresh layers of detail on both well-known and obscure missions (e.g. *Science and Space* on the Astron and Prognoz 9 missions). One of the last and the most intriguing in this collection was *The USSR in Space in 2005*, which was a prospectus of future space stations and missions to Mars.

Odd bits of news appeared in unusual places. For example, the story of how Pyotr Dolgov died testing the Vostok ejection system from a stratospheric balloon appeared in *Moscow News*, an English-language magazine circulated to tourists with probably quite a limited production run ('Leap from the stratosphere', 1988, Vol.24).

Listening to Radio Moscow

One of the most valuable sources was Radio Moscow, which broadcast in the shortwave on the 19, 25, 31, 41 and 49-metre bands. The quality of reception varied with atmospheric conditions, ranging from the crystal clear to the inaudible, in some case over the course of a single broadcast. There were actually two services, one for Britain and Ireland and the other for North America – for some reason, the

USSR Academy
of Sciences

V. P. Glushko ⸱17 IAN 1980

Rocket
Engines
GDL-OKB

2.3: Valentin Glushko got to publish in his own name.

American broadcasts were preceded by stirring martial music taken from the US Air Force band. News on spaceflight was carried on the hourly news broadcasts, on the magazine programme *People and Events*, and on the feature programme *Science and Engineering* that was hosted by Boris Belitsky, who had been Yuri Gagarin's translator during his visit to Britain in July 1961. The news reports, for example, gave daily accounts of the long-duration missions Soyuz 9 and Soyuz 11 which I would either annotate or else tape-record if possible. My first account of the Soviet space programme, *Race into Space* (1988) was largely based on these tape recordings and handwritten notes – a basic but effective record.

Radio Moscow provided reports on the Lunokhod operations on the Moon and the Venera and Mars missions. *Science and Engineering* interviewed scientists about the results of missions. *People and Events* routinely interviewed cosmonauts (Konstantin Feoktistov being a favourite), scientists (Mstislav Keldysh also a favourite), designers and engineers whose significance or role was rarely explained – some became better known in later years but others remained obscure. Sometimes, information leaked out that was probably not intended to. Radio Moscow once reported that Gherman Titov was testing a "spaceplane", the first admission that such a project existed (probably a reference to the Spiral programme). When Radio Moscow was inaudible, information could be picked up from some of the socialist bloc stations, of which Radio Prague had the best transmitter. In 1971 Radio Prague jumped the gun on the idea of socialist bloc cosmonauts flying to Soviet orbital stations, long before any such cosmonauts were selected in 1976 (Radio Prague, 13 March 1971).

Rewards and hazards

Following events on Radio Moscow and in *Soviet Weekly* was both rewarding and hazardous. At the rewarding end, Radio Moscow's reporting of the entry into Mars orbit of Mars 2 and 3 in 1971 told us that the Russians had developed autonomous astro-navigation years ahead of anyone else. The scientific results of the Venera, Mars and Luna missions of the 1970s were narrated in detail that was accessible to non-scientists. Radio Moscow gave out details on missions which were otherwise little reported elsewhere, like lunar orbiters Luna 19 and 22 and on individual Cosmos missions. *Soviet Weekly* published pictures that were rarely picked up in the Western press, such as human interest pictures of cosmonauts, profiles of the tracking fleet (16 January 1971 and 22 July 1972), Venera 5 tracked in the night sky (18 January 1969), a profile of Grigor Gurzadyan, designer of the Orion telescope on Soyuz 13 (7 August 1976), a profile of the Cosmos programme (21 June 1969) and coverage of individual missions, like the biological Cosmos 605 (9 March 1974). *Soviet Weekly* published microscope images, diagrams and analysis of the first rocks returned from the Moon by the Russians (7 November 1970), images of Lunokhod roving over the lunar surface (28 November 1970) and rare pictures from Luna 19 (19 February 1972) and Luna 20 (4 March 1972).

They became a useful source of information of planned or intended missions – but if these did not occur then this made one vulnerable to being 'wrong'. *Soviet Weekly* outlined plans for crystal furnaces in Earth orbit ('Stanislav Khabarov: Industry in space', 19 February 1972) – although these did not fly until Salyut 5 four years later. Other missions were sketched, such as a mission to Jupiter which never happened, although we know it reached the concept stage and was named Tsiolkovsky (Radio Moscow, 11 February 1972). Both projected a series of lunar missions to follow the post-Apollo round of samplers, rovers and orbiters that concluded in 1976. We now know that Lunokhod 3 and another orbiter were built but cancelled.

In advance of manned flights, listeners were often treated to announcements about the value of orbital stations – this was certainly the case in the run-up to the launch of the first Salyut orbital space station, which was well flagged in advance. Normally a manned flight duly followed, but on the occasions when it did not (e.g. July 1972 and April 1973) this was a sure indication that something had gone wrong. In the case of Salyut 2, the warm-up announcements ceased abruptly on 14 April 1973, the day on which (we later found out) the station suffered an explosion.

When Luna 9 landed on the Moon, Radio Moscow announced it as "a major step toward the landing by men on the moon", something it would hardly have done had Russia been planning only a robotic programme of exploration. At the cautionary end of the scale, once the first Voskhod was in orbit Radio Moscow said that the mission would be a "prolonged flight". The three cosmonauts were back on the ground a day later, hardly a prolonged flight, sparking inevitable speculation that it had been cut short. Years later we learned that the announcement had been in error; the flight had been intended to last only one day.

NEWS MANAGEMENT

For the space sleuth, an inevitable challenge was to cope with a phrase which was probably not in vogue at that time, although the practice was abundantly present – news management. One of the unbreakable rules of Soviet news management was that failures could never be admitted (except in the case of the two Soyuz disasters, Soyuz 1 and Soyuz 11, which could hardly be denied). As a result space probes made completely successful flybys of the Moon and planets (when really they were trying to land). When Luna 8 crashed while trying to soft land, all systems had functioned normally at all stages, with "except for the final touchdown" being added almost as an afterthought. The *Sunday Times* managed to land one of the few blows on news management when it reported that, far from making a normal re-entry, Voskhod 2 had come down in the Urals, the speed of descent burning its aerials away and the cabin coming down over a thousand kilometres off-course, with the two cosmonauts spending a freezing night in a forest awaiting rescue ('Voskhod 2 "enveloped in flames"', 'Soviet cosmonauts had closest escape of all', and 'Soviet welcome team waited in vain', 21 March 1965). At the post-flight press conference in Moscow the following week, the cosmonauts claimed they were "delighted" to have been given the opportunity to make a successful test of the backup manual re-entry system. All this made it difficult to separate reality from Russian news management on the one hand and Western coverage that was at times over-suspicious. For the record it should be noted that this was not a purely Russian vice. In April 1970 Voice of America told its listeners that there had been a "problem" with the electrical system on Apollo 13 which might mean a "change in mission".

The crash of Luna 18 while attempting to land in the lunar highlands in September 1971 is recorded as the first formally admitted failure ("it was unlucky", said Radio Moscow; which it probably was). Nevertheless, the old habits lingered. When the two Soyuz 15 cosmonauts failed to dock with Salyut 3 and returned after only two days, *Soviet Weekly* corrected the record: they were "testing emergency landings by night". Perhaps the most notorious spin was on the Luna 15 mission, which circled the Moon at the same time as Apollo 11. This presented news management with a challenge. *Soviet Weekly* ran 'No mystery about Luna 15' penned by the pseudonymous 'Pyotr Petrov'. It was simply a new and highly manoeuvrable lunar probe that happened to coincide with Apollo 11.

When Soyuz 9 returned to Earth, it was announced that the crew was in "perfect condition" after the 18-day flight. But that autumn, at the International Astronautical Federation (IAF), it emerged that they were so weak that they had to be carried from their capsule. *Soviet Weekly* wrote of the special facility used for convalescence, this having been designed to protect them from Earth bugs. In fact, they were testing out the Russian equivalent of NASA's Lunar Receiving Laboratory which was to prevent returning Apollo astronauts from infecting Earthlings with lunar epidemics; quite the opposite. Such misleading reports were quite rare, but they illustrated the importance of sifting fact from fiction in all this news management.

A working principle of Soviet news management was that whereas it rarely told an outright untruth it would readily mislead. A prime case of this was the Polyot series of manoeuvrable spacecraft, the two in the series being launched in November 1963 and April 1964 respectively. These were explained away as paving the way for Earth orbital rendezvous and space stations, setting up a false trail that this was the prime direction of the programme. In reality, Polyot was a military interceptor designed to close in on a hostile target, explode and destroy it.

The sheer volume of news coverage was often an important indicator. Although the objective of Luna 4, the first in a renewed series of probes, was not stated (April 1963) this mission to the Moon was accompanied by a fanfare of publicity of how cosmonauts would land there and establish bases. When it became apparent that the probe would miss the Moon, the official statement said that the purpose had always been to study to the Moon from close proximity as it flew past – and talk about bases quickly evaporated. But the volume and tone of the coverage in the first, optimistic 24 hours indicated that this was a new type of mission with an ambitious objective, in all likelihood a soft landing; as indeed was so. After a while, one learned to interpret events from the volume, tone and even U-turns in news coverage.

Missed opportunities

Sometimes, even managed news contained some important truths. When Soyuz 6, 7 and 8 failed to carry out their docking mission in October 1969, the following *Soviet Weekly* trumpeted 'They've brought space down to Earth!' with a full-page spread of the space observations carried out by the mission, devoted mainly to Earth resources observations. This was not the primary objective of the troika mission at all, but the article provided, for the first time, important detail on the development of space-based observations which it would have been easy to overlook. They were indeed the beginning of a programme subsequently extended by the Salyut orbiting stations.

NO MYSTERY ABOUT LUNA-15

says Pyotr Petrov

THERE is no mystery about Luna-15. It represents merely the successful latest in the series of Soviet lunar probes. And the information issued about it and its movements has been on the same lines as with all its predecessors.

Indeed, if it hadn't happened to coincide with the dramatic *Apollo* flight, it would hardly have received a mention in the British press.

Why did it coincide? It is inevitable that the USSR and the USA will make their moonshots at roughly the same times—because these are the only times when the Moon happens to be in the correct position for a visitor from Earth.

And, indeed, one earlier Soviet shot, Luna-12, was circling the Moon, busily sending back messages, for nearly five months: from August 25, 1966 to January 19, 1967.

During that period it circled the Moon 602 times—and there were no big newspaper headlines asking what it was up to!

What, then, has Luna-15 been doing? The answer is, a very great deal of very valuable research.

First, on its trip into Moon orbit, after launching on July 13, it carried out a wide range of checks into such matters as meteoritic materials and radiation to be found on the route.

Important and interesting in any context, this information is particularly vital for future flights—manned and unmanned—between Earth and Moon.

And so, on its 102-hour flight to the neighbourhood of the Moon, Luna-15 held 28 separate "report-back" sessions with the Ground Communications Centre.

The information it sent back is still being processed by Soviet research centres.

On July 17, Luna-15 went into the first of its orbits round the Moon.

Its tasks, then, included the continued detailed survey of the lunar surface begun by previous Soviet orbiting moon-probes: Luna-10, 11 and 12 in 1966, and Luna-14 last year.

Later, changes were made in this orbit, bringing the probe very much closer to the surface.

These orbits all had one thing in common, however, that they crossed the Moon at a very sharp angle to its Equator—unlike the *Apollo* capsule, which was almost following the line of the Equator.

The point of this becomes obvious after a moment's thought.

An equatorial orbit takes you round and round over virtually the same narrow stretch of the Moon.

This was necessary for the Apollo ships, as it enabled them to check the landing area again and again.

Even more important, it meant that the ship would continue passing over the landing site regularly every two hours, so making recovery of the landing vehicle possible.

Luna-15, however, had a different need. By orbiting at an angle, it takes advantage of the fact that the Moon is spinning beneath it—and so it crosses a different stretch of Moon every time round.

And, obviously, that's what you want from a survey ship.

Luna-15's second orbit, for instance, was at 126 degrees to the equator.

She circled the Moon once every 2 hr. 35 min., at heights ranging from about 59 to 138 miles.

Her third orbit was much lower: 68 miles to 10 miles up, circling once every 114 minutes. But the angle remained almost the same: 127 degrees.

After 52 orbits of the Moon, Luna-15 ended its work on July 21.

It represented an important advance on earlier moon-probes, in that it was able to vary its lunar orbit, and so land, if required, in many different areas of the lunar surface.

During its flight these automatic navigation systems were thoroughly tried out.

And so, Luna-15 ended her work. No mystery—just another

2.4: Soviet 'cover story' explaining the Luna 15 failure.

2.5: The 'Soyuz Seven' get star treatment in *Soviet Weekly*.

Soviet news management also missed some opportunities. For example, when Mars 3 soft-landed on the Red Planet on 2 December 1971 it began to transmit a photograph, until silenced by an electrical outage or sandstorm (we still do not know exactly). The image though, was not published until thirty years later when scientists in the Vernadsky institute were persuaded to replay the old signal and out came the first image from the surface of Mars. It was of poor visual quality but it showed the

Martian horizon (though American experts dispute this as 'noise'). Apparently, the original fell aesthetically short of what was considered publishable in 1971 and so was spiked.

Venus provided another example. The first photographs from the surface of the planet used fisheye lenses that severely curled the image. Although 'straightened' images were later published, they never appeared in popular external publications. In the same spirit, Leonid Ksanformaliti's recordings of lightning on Venus, detected by probes on its surface, were not released for over twenty years.

The Soviet Union was remarkably inconsistent in its approaches to photographs. Pictures from Luna 9 and Luna 13 taken on the surface of the Moon and long-range shots of Mars from Mars 3 were quickly published. On the other hand, although the Apollo 8 photograph of 'Earthrise' over the lunar horizon became one of the iconic images of the 20th century, similar images taken *earlier* by Zond 5 and Zond 6 were not exploited. Luna 12 photographed potential landing sites around the lunar equator but these were not published and may well have been lost. Later Zond images were of extraordinary density, and entire galleries of images were provided by Mars 4, Mars 5 and Fobos 2. Pictures of Venus taken by Venera 9 and Venera 10, the first probes to orbit around that planet, did not come to light for thirty years. From the point of view of sleuthing, their absence from the public record gave the incorrect impression of a lack of scientific results from these missions. The CIA instilled in its photo-analysts that "absence of evidence should never be interpreted as evidence of absence", which is an aphorism that we forget at our peril.

Visual observations

Leaving the written word or image aside for the moment, visual observations could still have an important role in interpreting the Soviet space programme. Predictions of satellite appearances were published in some British newspapers (*Daily Telegraph* and *Manchester Guardian*), which were in turn computed by the Science Research Council and distributed by Jodrell Bank. I obtained times for the Salyut space station, and saw it on several evenings in early September 1971 before it was deorbited on 15 October. When the second Salyut was launched on 3 April 1973, it was expected that a crew would follow within ten days. The fact that no crew followed suggested something was amiss; but Radio Moscow reassured me and anyone else tuning in that everything was alright. Obtaining the Jodrell Bank predictions, I was able to observe Salyut 2 fly over.

Unlike the steady brightness of its predecessor, Salyut 2 was uneven and flashing every 3 seconds. It was clearly tumbling, confirming suspicions that the mission was in trouble. When the flight ended, Radio Moscow did not use the word "successfully" but instead a coded and almost penitent phrase well-known to sleuths which signified disappointment: "the data from the mission will be used to inform the design of future space stations".

It was possible to follow subsequent Soyuz flights up to the Salyut stations, for they normally followed a 24-48 hour rendezvous pattern. The key challenge was to get the launch time from Radio Moscow and hope that this would give an overpass

```
                    SCIENCE RESEA( ) COUNCIL
               RADIO AND SPACE RESEARCH STATION
                    SATELLITE PREDICTIONS
          1973 17 A SALVUT 2        TIME = BST              MANCHESTER COORDINATES

             CO-ORDINATES OF SITE      53.5000 N.      -2.2500 E.

                                       LEAVES   HIGH  ELEV   ENTERS
          DATE        FROM      TO   RISES ECLIPSE  DEGS  IN   ECLIPSE  SETS
    BST    7  5 73    3  12.   3 14.    -    FSE    20  FSE    -    FSE       1
    BST    8  5 73    3  2.    3  3.    -    FSE    10  FSE    -    FSE       1
    BST    9  5 73    4  21.   4 24.    -    SSW    10  SSW    -    SSE       1
    BST   12  5 73   22 50.   22 52.   WSW    -     20  SW    SW     -        1
    BST   13  5 73   22 36.   22 40.   WSW    -     35  SE    SE     -        1
    BST   14  5 73   22 22.   22 27.   WSW    -     40  SSE   ESE    -        1
    BST   14  5 73   23 55.   23 56.    W     -     10  W     W      -        1
    BST   14  5 73   22  6.   22 13.   WSW    -     35  SE    ESE    -        1
    BST   15  5 73   23 41.   23 42.    W     -     10  W     W      -        1
    BST   16  5 73   23 26.   23 28.    W     -     15  WSW   WSW    -        1
```

2.6: These printed coordinates were used to track Salyut 2 over Ireland.

in early darkness over Ireland 7 hours 35 minutes later, with the Soyuz normally a couple of minutes distant behind the station itself. Salyut missions included quite a number of rendezvous operations that, for various reasons, went awry (e.g. Soyuz 15, 23, 25, 33, T8), and observing them visually from the ground was actually an effective means of assessing their progress.

My first success was Soyuz 18B chasing Salyut 4. The absence of Soyuz 18B in a separate orbit the following evening suggested they had docked successfully – even though Radio Moscow had not yet announced it. Reporting these observations, I was congratulated on my Einsteinian mathematical skills in figuring out how to predict the Soyuz 18B pass, coupled with a request for the formulae and algorithms involved in such a demanding, complex computation. I had to disappoint the sender by saying that if Baykonur was at 65°E and if an orbit moved 8.6° westward every hour, which we already knew, then this would put it over Dublin just over 7.5 hours later.

The Forum

A key event for followers of the Soviet space programme was the establishment by the British Interplanetary Society of its Soviet Space Forum, which usually took the form of a one-day event on the first weekend of June in London. This soon attracted a following far beyond Britain, welcoming participants from continental Europe and North America. In due course, "liaison officers" from embassies in London found a compelling need to work Saturdays in order to attend. These events enabled analysts to test and share information in a systematic manner. In the present information age in which news can be quickly shared across bulletin boards and blogs, it is easy to underestimate the importance of such a meeting place.

The following year, 1981, saw the publication of Jim Oberg's *Red Star in Orbit*. Oberg was a NASA mission controller who developed an interest in the Soviet space programme and had published in the British aviation periodical *Flight International* since 1974. He stubbornly believed that the Russians had always intended to reach the Moon first, and laid out his case in the classic article 'Russia meant to win the Moon Race' published in *Spaceflight* in 1975.

Oberg had devoted considerable attention to the importance of photographs in

2.7: Brian Harvey and Asif Siddiqi in discussion at the BIS Forum.

the record, tracking down pictures in which cosmonauts appeared, were brushed out, and then reappeared – rather like the more sinister political photographs of Stalin's time, immortalised in David King's *The Commissar Vanishes*. These cosmonaut pictures were important not just for the historical record, but because they explained missions planned but unflown. In Britain, Rex Hall was the leader of this programme of photo-analysis. It made him the world's foremost authority on the cosmonaut squad and the organiser of the Forum.

The photographs, and other events, highlighted many of the inconsistencies of the censorship with which we had dealt over the years. Yuri Gagarin's second backup, Grigori Nelyubov, who disappeared and reappeared in photographs of Yuri Gagarin en route to the launch pad, had been named in Yevgeni Riabchikov's *Russians in Space* published by Novosti many years earlier (1971). Hiding in plain view was a defining feature of Soviet news management. For example, Valentin Glushko was quoted frequently in *Soviet Weekly*, but his significance was never explained (he was the chief designer from 1974 to 1989, something never directly stated). Indeed, his book *Development of Rocketry and Space Technology in the USSR* (1973) gave us his biographical details up to 1933 but said nothing of his work during the intervening forty years. *Soviet Weekly* carried a lengthy article to mark the launching of Venera 7 ('Why another space probe is going to Venus', 19 September 1970) by-lined by a Dr

Mikhail Marov, but did not explain that Marov was their top planetary analyst. When Mikhail Tikhonravov died four years later he merited only a single paragraph, but as the person who led the team that designed Sputnik and who conceived the first lunar probes he was, after Korolev, the leading influence on the programme in the 1950s.

I even came across a personal case of air-brushing. An official photograph of the cosmonaut squad was published in 1969, but one of the cosmonauts, Boris Volynov, was absent from this group picture for no sinister reason whatever. When the picture was resupplied for my book *Race into Space* many years later, he was, in the interests of completeness, reinserted in the back row on the right!

One had to be fairly eagle-eyed to spot these oddities in photographs, but they could be unexpectedly informative. *The USSR in Space 2005* included an illustration of the Mir space station – but there was something odd about it since all the modules around the node were small, not the large modules that were actually employed. We later found out that the publishers had accidentally issued the original Mir design. For good measure, the book also included the prospective post-Mir orbital station which was intended to match the American Freedom space station, but we did not know that at the time either.

Designers and cosmodromes

The BIS was the place where we learned in 1983 of the release of a Russian-language book edited by Mstislav Keldysh, *The Creative Legacy of Sergei Korolev*. It was a limited edition and out of print within hours: no wonder, for it contained the design of the first lunar-landing plan, the Soyuz Complex. Copies eventually found their way to Europe over the following years. Not only did *The Creative Legacy* unveil important early designs, but it marked the beginning of our realisation of the importance of the Soviet design bureau, exemplified by an exposition to the Forum the following year by Claude Wachtel of the different and often rival design bureaus – a key moment in our understanding of the programme. Following this strand of analysis, US Air Force officer Bill Barry wrote the definitive doctoral thesis on the design bureaus and their fundamental role in shaping the evolution of the programme (*Missile design bureaus and Soviet manned space policy, 1953-70*, 1996).

The BIS was also significant for another form of analysis, that of the two-line element. British analyst Phil Clark was perhaps the most skilled mathematician at the Forum and he received regularly updated datasets on the orbits of Soviet satellites – launch time, apogee, perigee, inclination, period in orbit, equator crossing. Working from the axiom that particular orbits were selected for the purpose of a mission, it was possible to categorise all satellites in terms of their roles: photo-reconnaissance, electronic intelligence, navigation, and so on. It was also possible to infer clues as to the launcher used, and missions that failed or ended up in the wrong orbit. Clark has since been able to explain the mission of almost every Cosmos satellite (a vast undertaking for a series that has over 2,500 members). But back then a number of entries defied explanation and became the subject of his classic paper 'Obscure Cosmos Missions' (1993). Clark's techniques for analysing the orbits of the Soviet space programme are now being put to good use in explaining the unusual manoeuvres of Chinese satellites.

The diverse interests of those at the Forum meant it was possible to piece together detail on a variety of aspects of the Soviet space programme, as this list shows:

'Nuclear-powered satellites', by Charles Vick (1983)
'Launch failures', by Phil Clark (1985)
'The military Salyut design', by Neville Kidger (1986)
'Nauka modules', by Joel Powell (1987)
'Rocket engines', by Phil Clark (1987)
'Signals from the Mir space station', by John Branegan (1989)
'Vladimir Vernadsky, 1863-1945', by Anders Hansson (1990)
'The balloon programme', by Dave Shayler (1995)
'Analysis of the N-1 launcher', by Berry Sanders (1996)
'Sergei Korolev', by James Harford (1997)
'Arkon', by Geoff Perry (1998)
'Science on Mir', by Andy Salmon (1999)

None of these authors were full-time journalists on the Soviet (or any other) space programme and all had 'day jobs'. These amateurs filled a gap caused by the scarcity of full-time professional space writers. The daily newspapers were dependent on their Moscow correspondents, who had many other interests and were soon transferred to other assignments. During this period, only a few periodicals covered spaceflight in significant detail: *Aviation Week* (in the US), *Air & Cosmos* (in France) and *Flight International* (in Britain), which normally had a three or four page section devoted to spaceflight. In particular, *Flight* demonstrated what could be achieved by intelligent professional reporting by people who knew their field well. It used its resources and reputation to gain access to information beyond the reach of most amateurs. *Flight* took the initiative to ferret out information, and had a sharp eye for new pictures or diagrams of Soviet space equipment. It was often the first to publish fine images of Cosmos or Intercosmos satellites, and it once tracked down a wasp-shaped aerostat to explore the Venusian atmosphere (6 February 1982). For 'Glimpses of Voskhod' (14 September 1965), *Flight* asked Sovexportfilm for a special viewing of a film of the Voskhod 2 mission, extracts of which it then published. Later, it used fleeting images from the film *Steep Road into Space* to reconstruct a remarkably accurate impression of the Proton launch vehicle (16 August 1973). *Flight* published lengthy reports from the International Astronautical Congresses. Its reports on individual missions were classics (e.g. 'Soyuz 17: the great comeback', 20 February 1975). It also published detailed analyses, some of which were not by-lined (e.g. 'Portrait of Baikonur', 12 June 1975). But others were landmark events, such as Geoffrey Perry's 'Looking Down on the Middle East War' (21 February 1984), 'Cosmos at 74°' (30 November 1972), David Baker's 'Killer Satellites' (15 October 1972) and Jim Oberg's 'The Soviet Cosmonaut Group' (16 August 1973).

Interpretation

The overall challenge to the amateur was not just to collect the information, but to analyse it. The epic task, historically, was to determine whether the Soviet Union was truly trying to beat the Americans to a landing on the Moon in 1968-1969. Sleuths were forced to assess evidence from a broad range of sources: what the hardware told us (ambiguous), what the Russians told us (which also pointed both ways) and what the Americans told us (the CIA was absolutely positive in pointing to a Moon Race). Interpreters took diametrically opposite views. Once Apollo 8 orbited the Moon in December 1968, Soviet statements about going to the Moon became more muted and when Soyuz 4 and Soyuz 5 docked in Earth orbit and exchanged crews, the BBC told its listeners:

> These launches indicate that the Russians are moving on an entirely different track from the Americans, establishing platforms in Earth orbit.

This line was followed by all the British papers of the time, except one, the *Daily Express*, which broke ranks and ran this headline:

> *Moon race! Russia out to beat US*. Russia is going flat out to get a man on the moon before America (16 January 1969).

The hardware evidence was ambiguous. The Zond 5 around-the-Moon mission in September 1968 could, plausibly, be explained as part of a programme of automated lunar exploration (e.g. detailed photography of the far side, and the Zond maps were indeed superb) – but the fact that Jodrell Bank picked up a transmission of a human voice reading borsht menus indicated a more ambitious purpose. Assessing Soviet technical capacity was an inherently difficult task, but Western experts consistently underestimated Soviet capacity. A typical example was Luna 15, launched several days ahead of Apollo 11. There was much speculation as to its purpose, with some suggestions that it was a sample recovery mission (which indeed it was, and not even the first attempt because an earlier probe had been lost during launch). Britain's *Daily Telegraph* would not have been alone in this assessment:

> It is not thought that the Russians can make Luna 15 land and bring back samples. The technical complexities are thought to be too great ('Russia puts up her own moonship', 14 July 1969).

After Apollo 11, the Soviet Union presented manned Earth orbiting stations as the main line of space development, complemented by a programme of automated lunar exploration, even though the manned lunar programme was not actually cancelled for several years yet. News management examples in *Soviet Weekly* of the former were in 'Manned space stations – the next step' (27 September 1969) and 'Coming soon: a manned space station' (15 November 1969) and of the latter in 'It's the decade of the space robot' (3 October 1970).

It was generally assumed that Soviet technology was simply not fit for purpose. Not until many years later did we learn that engineers had tested out in Earth orbit their counterpart of the Apollo command and service module (the LOK) and over four missions tested out their lunar module (the LK) to perfection. Although the N-1 launcher which let them down was ridiculed for its complex systems ("a plumber's nightmare"), its engines, stored for many years, came to power the new American commercial rockets that began to fly in 2012. Its computer, the S-530, the same as used on the LK and the LOK, could relay data from the ascending N-1 at 9.6GB/sec; not bad for 1972. Soviet computers were widely believed to be inferior, but this was because they followed a different developmental path and looked different. The high-precision return paths from the Moon of the Zonds (1968) and the sample recovery missions (1970-1976), as well as the development of the first astro-navigation system for Mars missions (1971) were adequate evidence of high standards of mathematics, computing, tracking and control.

Back in the USSR

Eventually, it became possible to see space exhibits in the Soviet Union itself, the main venue being the VDNK, which was the exhibition of economic and scientific achievements in Moscow. The first surprise for me, in 1984, was to see the number of rocket motors exhibited there, including some that had not previously been reported in the West, such as the RD-301 tripropellant engine. The second, in 1988, was an entirely new satellite that had never been seen before. *What is that?* I asked myself, prior to snapping pictures of it from every possible angle. It proved to be an Okean maritime observation satellite.

2.8: Author's photograph of the Okean satellite on display in Moscow.

Not far from the entrance of the VDNK was a new, poorly signposted small museum underneath the Tsiolkovsky memorial. This had only four or five exhibits (it has since been turned into the magnificent, three-floor museum of cosmonautics) with pride of place given to the Mars 3 lander, which had not at that stage been revealed. The problem though, was that it was guarded by a formidable-looking elderly woman janitor who could have been a war veteran who fought at the front, and she scowled at me when I produced my camera to make it clear that taking pictures could result in an interview in the Lubianka. On the subject of which, next door to the Lubianka in Dzhershinsky square was Moscow's famous children's store, *Detsky Mir* (*Children's World*). This had a plastic lunar base toy which I bought for my four-year old. Only many years later was it apparent that – child's toy or not – it was based on one of the real moonbase designs of the 1970s. Hiding in plain view, again. Years later, strange objects continued to turn up, such as a Mars tractor built in the 1970s adorning the basement of the Institute for Space Research in Moscow.

GLASNOST

By this stage, *glasnost* ('openness') had begun to work its way through the space community. The first intimation of this came in late May 1985 when, conveniently just ahead of the annual BIS forum, Radio Moscow announced the forthcoming crew to the orbiting station Salyut 7, Vladimir Dzhanibekov and Viktor Savinyikh, who were launched a few days later. The commissioning crew for the Mir space station, Leonid Kizim and Vladimir Solovyov was not only announced in advance but the actual launch was covered live on Radio Moscow on the shortwave. From that point on, upcoming missions were announced weeks in advance, and later months ahead. Other hitherto obscured parts of the space programme were at last brought out into the open. For example, film was released for the first time showing a Proton rocket launching a VeGa probe to Venus in December 1984. This was, to say the least, long overdue, because the Proton was not a military rocket.

Mishin speaks out

The main impact of *glasnost* was on the history of the Soviet space programme. Cosmonautics day 1986 marked the publication in *Izvestia* of the first of a four-part series on the first cosmonaut squad, identifying their names and featuring their full picture for the first time. A few years later we began to get the true story of the Moon Race and how the Soviet Union had desperately attempted to beat the Americans to the Moon. Former chief designer Vasily Mishin gave details, informally, to Western interviewers and when nothing bad happened to him as a result, others followed suit. Although the technical details of the lunar programme (the N-1 launcher, the LOK lunar orbiter and the LK lunar lander) continued to give technical students many happy hours of analysis, the most significant aspect of the revelations concerned the rivalry between the designers and their institutes, and how their analysts catastrophically underestimated the importance of the Kennedy speech of 25 May 1961 (because the programme that he started seemed literally incredible) with the result that the Soviet Union delayed until August 1964 its decision to contest Apollo.

Delayed response to blame

Instead of a command economy, we saw a chaotic system of political decision-making, directives revisited and revised, decisions overruled, "socialist competition" running out of control, and a process in which personality was all-important. It meant we had to revise our history of the Moon Race completely: the Soviet Union lost not because of technical inability (far from it) but managerial incompetence. Individual decisions and factors came to play a disproportionate role in the Soviet space programme, as witnessed by the climax of the round-the-Moon effort in December 1968. Although the Soviet Union had the technical window to go around the Moon on 7-9 December, two weeks ahead of Apollo 8, no such launch occurred. The principal reason was the caution of the chief designer, Vasily Mishin (he was unknown to us at the time) and a conservative approach to safety (quite contrary to popular Western impressions of a Russian recklessness with human life).

My first book, *Race into Space* (1988), was written without the benefit of most of the *glasnost* revelations, which had yet to come, but they informed my second book *The New Russian Space Programme* in 1996. Apart from some details which proved to be wide of the mark, most proved to be reasonably accurate thanks largely to the BIS Forum where we could all test our interpretations against each other and at least ensure that the fragmentary evidence base was reviewed carefully and that any errors were well-informed errors.

MEETING COSMONAUTS AND SCIENTISTS

A further consequence of *glasnost* was that Russian space designers, cosmonauts and scientists had more freedom to travel and talk about their missions and experiences. For the historian, journalist or amateur sleuth, there is no substitute for listening to people relate their experiences in their own words. After his last mission, cosmonaut Georgi Grechko travelled to Dublin to speak of his three flights on the Salyut space stations, giving us a detailed perspective of living and working in space as well as ascent to orbit and the return through fiery re-entry.

Grechko was followed by Konstantin Gringauz, who arrived on the invitation of Susan McKenna Lawlor, an Irish scientist who flew equipment on the Fobos mission to Mars in 1988-1989 which determined the radiation level in Mars orbit. I had never heard of Konstantin Gringauz before (though it transpired that I really ought to have) and he spoke of his experience in designing and building Sputnik – indeed he was the last person to hold it in his hands before it was dispatched heavenward. This modest man said nothing of his role as one of the leading scientists of their space programme, a point I was to appreciate later. He died not long afterward.

2.9: Konstantin Gringauz, a modest man more appreciated after his death.

Later, I endeavoured to get to as many talks as possible given by cosmonauts, such as Alexander Volkov, Yuri Usachev, Mikhail Tyurin, Alexander Kaleri, for each one had different things to say and unique and different insights. Dr Valeri Polyakov was especially important, for he flew what is still the longest duration mission, 438 days, and his reflections on that experience are still relevant to us today.

End of an era?

By the new century, the work of the amateur space sleuth was largely over in terms of the Soviet space programme, but a new door opened because many of the same dramas were being replayed in China. In recognition of this, the BIS Forum was renamed the Soviet and Chinese Forum. Embassy liaison officers could safely take Saturdays off again. Several of the designers, like Boris Chertok, had told their stories in memoirs. Diaries had been published, notably that of the cosmonaut squad commander Nikolai Kamanin, translated by Bart Hendrickx. Rex Hall and Bert Vis made annual, lengthy visits to Star Town to complete our knowledge of the cosmonaut squad, while group tours to Baykonur became commonplace.

The archives were opened, at least partially, enabling Asif Siddiqi to make his definitive contribution to the Soviet side of the Moon Race – first with *Challenge to Apollo* and thereafter his captivating prequel *The Red Rockets' Glare*. Furthermore, the Energiya, Khrunichev, Yuzhnoye and Kozlov design bureaus published company histories. Russian historians and journalists established a regular monthly magazine, *Novosti Kosmonautiki*, which is now recognised to be amongst the best in the world. Launch manifests were issued long ahead of time and new cosmonauts were revealed in advance. We arrived at the situation – which would have been unimaginable forty years earlier – of knowing more about Russian military satellites than American ones, whose orbits, contrary to international law, were not even announced.

Space science

One aspect of the Soviet-Russian space programme which had not come fully into the light of day was space science – in other words, the outcomes of their space missions. The conventional wisdom was that they did not publish their scientific results. There was something about this that did not ring true. The person who broke this narrative was Kenneth Gatland, who served as editor of *Spaceflight* magazine, was a regular contributor to *Flight International,* and wrote a series of books on space exploration published by Blandford, notably *Robot Explorers* (1972) and *Manned Spacecraft* (1972). He had been able to lay his hands on some of the results of Soviet scientific space missions and sourced these at the annual International Astronautical Congress (IAC) and the Committee on Space Research (COSPAR). It was he who broke the story in *Flight International* of the poor condition of Andrian Nikolayev and Vitaly Sevastyanov after their Soyuz 9 mission: he had gone to the conferences, attended the presentations and spoken to the presenters, so his sources were impeccable.

Another straw in the wind that we had not heard the full story about came with the publication by the University of Arizona in 1992 of its mammoth book *Mars*, which included a chapter on Soviet missions to Mars in the 1970s co-written by planetary scientist Vasili Moroz. This was quite contrary to the received wisdom that all these missions had failed (indeed, when the Fobos-Grunt mission failed to leave Earth orbit in 2011, this mantra was drearily regurgitated). The chapter in *Mars* described the descent of Mars 6 and how it had relayed real-time data throughout, giving us the first-ever near-surface measurements from the planet in 1974, two years before the Viking landers. Further details of the descent were then given at an unlikely event: a workshop on descent trajectory analysis held in Lisbon, Portugal, eleven years later. Had more been achieved than we realised? Radio Moscow and *Soviet Weekly* barely mentioned the arrival of the Mars 4, 5, 6 and 7 probes that year, so the official Soviet position at the time was consistent with the Western impression of total failure. Was this some perverse form of news management? What was the real story?

Some other sources hinted that the space science treasure trove was substantial. The periodical *Zemlya i Vselennaya* (*Earth and the Universe*) ran stories about the successful Granat astronomical observatory and, remarkably, of the construction of miniature 'holodecks' on the Salyut 6 and 7 orbital stations. Anders Hansson, who presented exotic papers at the BIS, spoke of science experiments designed to detect earthquakes from orbit, growing wheat in orbital stations, and on the effects of free radicals on rats and monkeys. Soviet scientific papers presented at NASA's annual lunar and planetary science conference hinted at more. NASA convened the first of these events in January 1970, a serious, worthy attempt to bring together the learning from both the Apollo rock samples and the scientific instruments left on the surface of the Moon, in an effort to inform the wider scientific community. Soviet scientists were invited to these gatherings, which they attended the following year, when they had something to bring to the party (Luna 16's rock samples). The press reported on these conferences, but never, to my recollection, on the Soviet contributions which were often substantial. The scientific haul from the core sample of Luna 24, coming four years after the last Apollo landing, was so substantial that the papers presented to the 1978 conference were issued as a book called *The view from Luna 24*. Also in the United States, Don Mitchell has digitised and reprocessed the Russian planetary and lunar imagery archive, made an inventory of spacecraft instruments and identified their designers.

COSPAR

To return to the IAF and COSPAR, I found a handful of their papers (nicely bound and mainly from the 1960s) at the BIS library in London. Its early collection reflected what people like Kenneth Gatland had collected and what space agencies considered fit to send the BIS at that time, but in the spirit of true librarianship the papers had been kept, sorted, and well looked after over the years. Sure enough, they contained papers on the early lunar, interplanetary and Earth-orbiting Cosmos missions written by scientists who were little heard of, including my now-departed friend, Konstantin Gringauz. Perhaps the most remarkable find in the BIS library was what are called the NASA TTF papers (TT stands for technical translation). Apparently, in its early days, NASA decided that it should translate the main scientific and related papers published by other countries, principally Russia, but the agency also kept an eye on the Japanese. In forty years of following spaceflight, I had never seen a reference anywhere to the TTF papers, but they were an extraordinary collection of translated Soviet announcements, journal articles, press pieces and conference proceedings. For the investigator of Soviet space science, these were a Godsend. I have no knowledge of the degree to which these papers were used, either within NASA or the American space community at the time – if at all. The results of these space missions were not just of interest in themselves, they also gave considerable insights into the technical capabilities of Soviet instrumentation and spacecraft.

The problem was that this trail of IAF and COSPAR papers evaporated in the 1970s when the IAF ceased to systematically keep papers. The printed TTF papers died out in 1974, but NASA subsequently and helpfully digitised the whole series into PDF format, best searched by author names. For the historian, though, NASA performed a huge and little-acknowledged public service.

This left me with two important information banks to track down: the remaining COSPAR papers; and the in-house publication of Soviet space science, the existence of which was apparent from the TTF papers. COSPAR, which dated to the period of Sputnik and the late 1950s, appeared to be a United Nations body but it was not. It was intended that it should be, but the scientists of the time decided that if they were to meet under the auspices of the UN, the process would become bureaucratised and politicised, hence they made it a stand-alone body. Furthermore, the scientists of the USSR and USA took the view that its location should be with neither of them, but a more neutral European location. Since Britain was considered at that time the most likely leader of the European space community, the initial proposal was for location in London. But the French established their own space agency and invited COSPAR there. It is now housed within the offices of the French space agency CNES (Centre National d'Études Spatiales). The initial annual volume of COSPAR papers in the 1960s has grown beyond all recognition, with thematic series for all aspects of space research, but there, in the COSPAR papers in Paris, were the outcomes of decades of Soviet publishing in space science.

Hidden journals

This left the in-house journal of space science which started at the beginning of the Space Age. Soviet science was, in the literal sense of word, highly institutionalised by adopting the European practice of being based in institutes, rather than the American practice of universities. It was nothing if not systematic, the disciplines starting with a periodical restatement of their knowledge base. And with this went a commitment to dissemination. Therefore the establishment of a journal of space science was only to be expected. This duly appeared as *Космически Исследоватл* (Kosmicheski Issledovatl in Russian, published for the convenience of English-speaking readers as *Space Research*). But where was it? And the early editions in particular? A colleague at the BIS Forum recommended trying the Science Museum Library. It transpired that the library had long since run out of space in London and had transferred much of its collection to an old wartime transport base on a plateau above Swindon in Wiltshire, on the railway line between Swansea and London. The old hangars had been air-sealed in order to preserve the texts at the right temperature and humidity. Reaching them involved crossing a series of long, wide runways and the only thing missing was for a lumbering World War II transport or bomber to come in to land, like in the film *The Memphis Belle*.

Between the COSPAR and the Swindon collections, I found the assembled papers of Soviet and Russian space science and it was a realm that had been little explored before. To convey a sense of how little explored it was, my former college, Dublin University (Trinity College), had an original version of the 1959 interpretative map of the Moon which was drawn up after the flight of the Automatic Interplanetary Station later renamed Luna 3. It was in absolutely crisp, untouched condition, the pages still sticking together ever so slightly, and the mark on the inside cover page indicated that it had not been opened since its arrival in Dublin fifty years earlier – I was the first person ever to have requested to see it.

In the event, between *Space Research*, the COSPAR papers, and other available documentation, it was possible to compile a substantial record of Soviet and Russian space science. In Earth orbit, Cosmos satellites had measured air density, Prognoz observatories had made detailed studies of the Sun over twenty years, while Elektron missions mapped Earth's magnetic fields. Soviet space science had invested hugely in magnetospheric research, understandable given that country's northerly latitude. Mars 6 had indeed made a profile of the descent down to the surface of Mars, while Mars 4 made a map as it flew past. Venera 15 and Venera 16 sent back beautifully contoured radar maps of Venus, while the VeGa missions gave us a close-in analysis of Halley's Comet and Fobos 2 inspected Phobos, the larger of Mars' two small satellites. The magnetic fields of Venus and Mars were characterised. A wealth of astronomical data came back from Granat, supplemented by Astron and Kvant. There was also plenty that was exotic. Cosmos 1809 and other missions established a relationship between the occurrence of earthquakes and magnetic distortions hours beforehand, suggesting it would be possible to establish a predictive service. Ionozond picked up the plume from the Three Mile Island nuclear disaster in America. Nikolai Basov managed to get a molecular oscillator into orbit on Cosmos 97 and 145 in order to test theories of time and relativity. Leningrad physicist Boris Konstantinov had detectors installed on Cosmos 135 and 167 to find antimatter and the remnants of comet showers in Earth orbit.

Not only that, but the research revealed a world of personalities quite unknown to the West. Konstantin Gringauz was first to encounter the solar wind when the First Cosmic Ship, later renamed Luna 1, sailed into solar orbit in 1959. Yuri Lipsky was the driving force behind the maps and atlases of the Moon, Mars and Venus. Gavril Tikhov was an early analyst of the conditions for life on Mars (and less probably on Venus). Iosif Gitelson developed the closed-cycle life support systems that will one day make interplanetary human travel possible. Yuri Galperin pioneered iono-spheric research. Vasili Moroz revised our theories of the exploration of the planets based on the results of the Venera and Mars probes. Sergei Vernov was the man who, with his American colleague James Van Allen, built our knowledge of the radiation belts. All these people were historical personalities in their own right, each with a story to tell of the evolution of not just space science, but science in the Soviet Union and the politics thereof.

Visual record

Another discovery was a visual one. A recurrent feature of Soviet publications was the often poor visual quality of the images presented, literally in sharp contrast to the full-colour American presentations published in popular outlets such as *National Geographic*. This led us to the false conclusion that Soviet photographic equipment was poor, an assumption laid to rest by the research of Don Mitchell. Lacking the commercial imperatives of Western publishers to seek ever better production values, the Russians often made reproductions of reproductions, becoming grainier each time. But it transpired that the Soviet Union had handed over a set of originals to NASA as part of their programme for lunar and planetary cooperation. Canadian mapper Philip J. Stooke managed to track down more originals in Moscow, which he published in his *International atlas of lunar exploration*. Perhaps inspired by this,

ПОВЕРХНОСТЬ МАРСА

ИЗДАТЕЛЬСТВО «НАУКА»

2.10: 'The surface of Mars' contained a beautiful fold-out map based on the Mars 4 flyby.

the Vernadsky institute later uploaded the entire Lunokhod archive onto the Internet.

My study also shed further light on the factors that help or hinder the space sleuth. The first lesson was that much of the information was actually there, but we weren't looking in the right places. Few of us had managed to attend either the IAC events or the COSPAR ones – but in our defence, most of us had day jobs and finding the time as well as the resources to travel posed a challenge. And in addition to *Kosmicheski Issledovatl/Space Research*, the old Soviet Union published its scientific results quite extensively. Some of these texts can still be obtained via second-hand book outlets. The main publisher was Nauka Press ('nauk' is Russian for 'science'). The range of subjects was extensive. Notable examples were *Поверхность Марса* (*Poverkhnost Marsa, The surface of Mars*), with a pull-out map of Mars at the end based on the Mars 4 photographs – which is now framed and displayed on my wall – and Leonid Ksanformaliti's *Planeta Venera* (*Planet Venus*) released in 1985. The format was reminiscent of British Victorian natural science publications, with pull-out sections that required gentle handling. The books on Luna 9 and Luna 13 included the original photographs, not the ever-degraded reproductions we later saw. Even if universities and libraries in the Anglo-American world were uninterested in the publications of Nauka press, we sleuths should have found our way to them sooner than we did. We should also have understood that the role of primary investigator (PI), standard for Western scientific space missions, was paralleled on the Soviet side with the 'glavny experimentator' (general or senior experimenter), and direct contact with them would have been our route to more knowledge, sooner. Whenever I did manage to contact PIs, which is now more easily done via the Internet, they have always been obliging with the provision of information, papers and permissions to publish.

The parallel universes

A basic problem arising from the rivalry between the United States and Soviet Union in the Space Race was that two parallel universes of knowledge grew up. This was not the intent of the scientists of the late 1950s. For example, not only were Sergei Vernov and James Van Allen good friends, but each made lecture tours of the other's country. By the mid-1960s though, this cordial atmosphere had withered and, as one commentator put it, if an American applied for a research grant from the National Science Foundation and included a collaborative paper with a Soviet scientist on his CV, he was more likely to receive a visit from the Federal Bureau of Investigation than a grant. Consequently, despite NASA's best efforts with the TTF series, Soviet achievements in space science were not known, ignored or forgotten. This led to the presumption that they did not publish their results, or, if they did, they were so poor as not to be worth publishing in major journals. On occasion this became comical, as when American scientists discovered a third radiation belt around the Earth, only to be reminded by their Soviet opposite numbers that they had discovered it many years earlier – and publicised it at the time. But nobody had read it.

A counter-narrative sprang up in which the only discoveries worth making were

Western ones. A quick look at the planetary and astronomy sections of university libraries illustrates the point well, for texts that reflect Russian research can rarely be found. This counter-narrative is not just an issue of irritating factual inaccuracies, but a worldview in which nothing outside one particular paradigm requires consideration. Two examples of the former are that the last lunar rocks to be returned to Earth were retrieved by Apollo 17 (not so: Luna 24); and that Magellan made the first radar map of Venus (not true: Venera 15 and Venera 16, although their coverage was partial). As recently as October 2011 *Spaceflight* reported that the American Mariner 2 probe to Venus discovered the solar wind in 1962 (not true: Gringauz' First Cosmic Ship in 1959). The search for water on the Moon is attributed to the Clementine and Lunar Prospector missions in the 1990s, but in 1977 a telescope on Salyut 5 made a scan of the Moon from Earth orbit that implied an abundance of water. The current axiom of the "warmer, wetter" Mars dates to Vasili Moroz, who proposed it in the 1970s.

Just as relying on supposedly authoritative information has its dangers, so too does not paying adequate attention to those with questionable reputations. Sleuths ignored the findings of the Central Intelligence Agency at their peril, for its original reports on failed Soviet launchings set the standard. During the late 1990s, President Clinton declassified many of its analyses of the early Soviet space programme (and some of the Chinese too). These included a detailed analysis of Soviet space science, its strengths and weaknesses. We know that the CIA was rigorous in its analysis, carefully using trigonometry to calculate the dimensions of Soviet launchers, but the CIA is less well known for the careful way in which it weighed evidence about Soviet scientific and technical capabilities. Most importantly, it never made the mistake of assuming that "because we do things this way, they must do the same". In *The case of the SS-6*, M. C. Wonus stated:

> The principal shortcomings of our analytical cycle did not result from mistakes in the interpretation of the available data, nor from deficiencies in the quality or the quantity of the data. Instead, error most frequently arose from attempts to relate Soviet technology directly to [our own]. It is now evident that this approach involves a dangerous assumption. Erroneous judgements reached by ignoring available intelligence because it gives answers inconsistent with 'our way of doing things' have unfortunately been common in the scientific intelligence field. Those who turned out to be wrong based their decisions on domestic logic rather than by objectively interpreting available intelligence information. An erroneous assumption over-emphasizing the importance of comparable US practices led the community astray. The Soviet approach can be radically different. The failure of the intelligence community to recognise that a multi-chambered engine was employed in the SS-6 was embarrassing.

Wonus was probably being very critical of his colleagues, since they got almost everything else right. Nevertheless, his exhortation should probably be on the advice list of any space amateur. Most of the CIA analysis was based on satellite imaging, supplemented by, some believe, human intelligence ('humint', i.e. spies). One of the most remarkable declassified documents is the minutes of the 1960 annual general meeting of the Soviet Academy of Sciences where there was, ironically in the light of everything said thus far, a plea to improve the production standards of dissemination for scientific papers in order to match the best in the West. How it leaked out is still a mystery.

CONCLUSIONS

Several important lessons emerge from the experience of sleuthing the Soviet space programme from the 1960s onward. Principal of these are:

- The requirement to use a combination of methods, ranging from information collection to visual observations, and from interviewing major personalities to attending conferences.
- To employ as wide a variety of sources as possible, from radio to booklets to books to the full range of what librarians call 'grey literature'.
- The need to recognise and interpret the subtleties of news management and in particular to distinguish between accurate statements, an economy of the truth, and misleading or even false statements; along with need to develop the skills to interpret news stories by volume, pitch and tone.
- To challenge the conventional narrative, in our case a strongly prevailing and politically informed force. This led us astray in so many ways, especially the assumption that the Soviet space programme was run as a command economy in which personalities were unimportant, or that scientific results were sparse.
- The dangers of underestimating the subject, in this case the technical abilities of the Soviet Union at a time when the conventional Western narrative was, in spite of its undisputed achievements, one of 'backwardness'.
- An awareness that unexpected sources can provide high-quality and accurate information, principally the CIA.
- The value of international collaborative forums, such as that created by the BIS, especially in the pre-Internet age.
- The importance of alertness to information appearing in unexpected places; of leaks; of the accidental supplying of stories, or designs when not intended.

Due to *glasnost* introduced by Mikhail Gorbachev and sustained by Boris Yeltsin, the main lines of the story of the Soviet space programme are now well known. It has entered a more familiar norm of conventional, mainstream history, where new details continue to emerge. Nevertheless, there are still obscure corners into which hopefully light will one day shine. And the techniques developed by amateurs in following the Soviet space programme can usefully be applied to the cosmic history of the country that invented the rocket, China. But that is another story.

3

The satellite trackers

by Sven Grahn

I was born in Stockholm, the capital of Sweden, less than a year after the end of the Second World War – which was just at the right time to catch the "space bug" when the Space Age began a decade later. The early Space Race between the United States and the Soviet Union occurred during my childhood, and I dropped my toy trains and airplanes to follow developments breathlessly.

During those early years I often read news stories about radio stations, such as the monitoring station of the Swedish telecommunications agency, picking up radio signals from Soviet space satellites – they even heard cosmonauts talking! Radio amateurs could also hear these signals on shortwave, so I reasoned it should not be impossible for me to hear them too.

LISTENING TO SHORTWAVES

I longed for a more 'hands on' contact with space technology after I had worked as a rocket assembly technician at the first Swedish sounding rocket base at Kronogård in northern Sweden. The rockets were a joint US-Swedish project during the summers of 1962-1964 to study noctilucent clouds. I got this fantastic job as a sixteen-year-old enthusiast because I was a member of the Swedish Interplanetary Society and it had arranged for some of its junior members to work as apprentices – and I really worked with the rockets themselves. In October 1964, at the age of eighteen, I entered the Royal Institute of Technology in Stockholm for the degree of master of engineering.

During that year I had also gotten in contact with a space tracking enthusiast Jan Jutander, who had been tracking satellites for several years from his parents' house in Tolarp outside the Swedish city of Jönköping. Jan's efforts were quite sophisticated and he had plenty of space to erect a variety of clever antennas. He convinced me that even I could listen to satellites, and so in October 1964 I bought a Lafayette HE-30 shortwave receiver for the (to me) "astronomical" sum of $90 in an

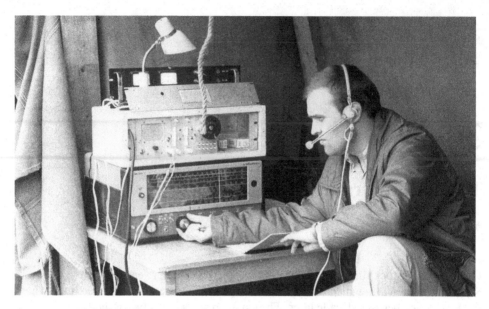

3.1: Sven Grahn tuning in to shortwave signals in the mid-1960s.

electronics store in downtown Stockholm. However, it lacked an important feature: an accurate tuning dial.

At the end of 1965, after a year of trying, I could finally tune the radio exactly to 19.995 MHz, which was the most common Soviet satellite frequency on shortwave. This was used by the type of spacecraft in which Yuri Gagarin had ridden into space, and since 1962 had been used by satellites in the 'Kosmos' series which remained in orbit for eight days and then returned to Earth. These were obviously intended for photographic reconnaissance, with the capsule returning the exposed film. The launches were announced in telegrams from the official Soviet news agency, TASS, and read out on Radio Moscow's English-language broadcasts, which I monitored.

Finally a bit of luck

On 7 January 1966 I came home at 1330 local time. Unusually, my father, Evert, was home from his job as purchasing manager of the newspaper *Dagens Nyheter* (*Today's News*). When I stepped through the door he told me that he had heard on a radio news broadcast an hour earlier that the Bochum Observatory in Germany had announced that they had picked up signals from a newly launched Soviet spacecraft. I switched on my HE-30 receiver and sat down with my textbooks as the radio hissed in the background. At about 1500 I heard a 'beep-beep' coming from the radio which I thought came from a satellite, and started to record the signals until they gradually faded out after seven minutes. They came back an hour and a half later – a typical period for a satellite travelling in low orbit. I was quite sure that the signals originated from Kosmos 104.

I had read two articles in the British aviation magazine *Flight* about a grammar school in the little town of Kettering in Northamptonshire that was listening in to satellites, so I wrote a letter to them and included a tape reel of my recorded signals. I got a prompt reply the following week – remember, this was in the days long before email, faxes and the like. In the letter Geoff Perry confirmed that I had indeed heard Kosmos 104, and that I should listen out for another new satellite that had just been launched. He also said that I ought to get a stopwatch to make the reception times in my monitoring reports more accurate. This exchange of letters was the beginning of a 34-year-long friendship with Geoff. I had become a member of a loosely-knit group of satellite listeners around the world.

The Kettering Group

Geoff Perry (1927-2000), the leader of the 'Kettering Group', was my mentor and constant letter, phone and email contact from 1966 until he passed away in January 2000.

The story of how he came to monitor satellite signals starts in his home town of Braintree in 1944 when he heard the double sonic booms from a German V2 missile that fell a few kilometres from his home. Geoff studied at the University of Reading and joined the Kettering Grammar School as a physics teacher in 1957. At a teachers' conference he heard signals from Sputnik 1 and was impressed by the clear Doppler shift in the frequency of the signal and he decided to use it in the classroom to explain the Doppler effect.

In order to make his own recording of a satellite signal Geoff sought the help of the chemistry teacher, Derek Slater. Early on a Monday morning in May 1960 they succeeded in recording signals on 19.995 MHz from Sputnik 4, which was the first test flight of the spacecraft design that would be used for Gagarin's Vostok. Later the same week, Geoff and Derek noticed that the over flights of Sputnik 4 were later than expected and concluded that its orbit had been raised. They were absolutely right – it was the result of firing the satellite's retrorocket in the wrong direction, which raised the orbit instead of lowering it into the atmosphere.

In the spring of 1962 the Soviet Union started launching satellites in its Kosmos series, many of which transmitted on the shortwave and were perfect 'mini projects' that could be used as a teaching aid to explain the principles of physics ranging from celestial mechanics to the propagation of radio waves. I suppose it was Geoff's way of making the subject of physics exciting.

Kettering, a town near the former steel works at Corby, is perhaps best known for the Weetabix breakfast cereal. The present town grew up in the 19th century with the development of the shoe industry. The research institute of the Shoe and Allied Trades Research Association (SATRA) had a computer which Geoff and Derek used to process their satellite observations. Sometimes they also used the computer of the Desborough Cooperative Wholesale Society corset factory!

International membership

The Kettering Group developed gradually. In the beginning, some pupils were given

3.2: Kettering Group members Sven Grahn and Robert Christy with the Vostok 1 capsule.

keys to the physics laboratory in order to enable them to make radio observations at weekends. The first pupils to make regular observations were the sixth-form girls from Kettering High School who came to the Grammar School (a boys' school) for physics, chemistry and mathematics. They sat in the empty physics laboratory during some of their free periods because there was nowhere else for them to go. That was the real start of pupil involvement.

Geoff's teaching methods were reported at teachers' conferences, and were also reported in newspapers thereby giving Geoff and Derek's operation recognition and useful contacts within the UK. In 1965 the aviation magazine *Flight* published a two-page article entitled 'Kettering's Cosmos Scholars', which attracted the attention of similarly inclined people like myself.

In the spring of 1966 I would just let the radio hiss away in the background while at home reading textbooks or doing other engineering student chores. A succession of photo-reconnaissance satellite in the Kosmos series were launched and I diligently listened to their signals. The satellites were launched into orbital planes inclined at an angle of 65° to the equator. This meant they traversed all areas of the globe between 65°S and 65°N, and therefore passed over Sweden regularly. The orbital altitude was typically 205-325 kilometres and the orbital period was 89.7 minutes.

These early recoverable reconnaissance satellites were a variant of the Vostok spacecraft and had the designation 'Zenit 2' (which I certainly did not know at the time). The first one to be placed into orbit was Kosmos 4 launched on 26 April 1962, and it remained in orbit for three days before the big spherical capsule carrying the cameras and exposed film came down by parachute. Both the flight duration and frequency of flights then increased gradually. On average there were 17 days between launches. Since the lifetime of each satellite was eight days the time when there was no Soviet photo-reconnaissance satellite in space was nine days. However, the tempo was so high that sometimes two satellites were in orbit simultaneously.

This furious launch rate was such that you really wonder what prompted it. Each launch was a major undertaking almost on the same scale as launching a cosmonaut. The only possible conclusion is that the need for reliable intelligence on the nuclear arsenal of the United States was extremely high, and that the international situation must have been tense. But the spring of 1966 was not, in my recollection, any worse than any other period during the years either directly before or after. So we are led to the conclusion that the launch rate shows how tense the geopolitical situation was in general during the height of the Cold War.

TK signals

Since I was new to satellite listening, I did not have any prejudices or preconceived notions and that helped me in my first 'discovery'. Here is how it happened.

When a reconnaissance satellite in the Kosmos series was preparing for landing after 127 circuits of the globe – just under eight days after launch – it transmitted signals as usual on 19.995 MHz but sometimes the frequency of the signal could be heard to shift suddenly when the retrorocket was ignited to start the descent phase. When the spacecraft started to be surrounded by ionised gas during atmospheric re-entry the signals quickly faded out. The transmitter on 19.995 MHz that we heard during the flight was carried in the service module that burned up in the atmosphere after separating from the large spherical descent capsule.

On Thursday 14 April 1966 it was time for Kosmos 114 to return to Earth after completing its mission. Both I and Geoff Perry were, independently of each other, monitoring the frequency-shift keyed signals from the satellite during its final orbit. In Stockholm the signals were heard at 0802.30–0810.08 local time. In Kettering, where Geoff listened to the signals in the physics laboratory a lesson was perhaps about to start, so he may have switched off the radio as soon as the normal signals disappeared as expected. I, on the other hand, a novice, left my radio switched on. At 0825 a slowly transmitted series of Morse characters appeared on 19,995 MHz. They were the letters TK, usually pronounced "dah dah-di-dah", and they were sent in a continuous stream that lasted until 12 o'clock.

In August 1966 I heard signals of the same type in connection with the landing of Kosmos 126. When we later collected observations from eleven Kosmos recoveries, we established the average time between the end of normal signals to the start of the TK signals to be 6 minutes and 45 seconds. The strength of the TK signals would drop suddenly after about eight minutes, most likely at the moment that the capsule touched down. The start of the TK signals probably represents the moment when the main parachute deployed. Thus the TK transmitter was placed in the capsule and its transmissions served as a beacon for locating the landing spot by using the Soviet network of "Krug" (Circle) direction-finding stations still in operation in Russia. Later we picked up other variants of these beacon signals from slightly different models of the basic Zenit spacecraft, known for obvious reasons as the TG, TF, and TL signals.

This shows the value of accurate log books and well-labelled recordings; the basic credo of electronic intelligence gathering!

Frequency-changing 'trick'

We also discovered something very peculiar that made it possible to predict when a new photo-reconnaissance Kosmos satellite would be launched – to within a day. We noticed that sometimes a photo-reconnaissance Kosmos would suddenly change its transmission frequency from 19.995 MHz to 19.990 MHz. It turned out that this was a clear sign that another such satellite would be launched within 24 hours. The new satellite would then appear using the older satellite's frequency of 19.995 MHz! One wonders why this was so necessary. It seems there was a rule that the latest satellite was supposed to transmit on 19.995 MHz. Since it was quite common that two photo-reconnaissance satellite were in orbit simultaneously, we could often make amazingly accurate predictions about new satellite launches.

The TK signals and this frequency-changing 'trick' were welcome discoveries but monitoring the photo-reconnaissance satellites was mostly routine and a way to hone our observation methods in preparation for more exciting space adventures like the flights of cosmonauts.

A Soviet cover story?

It may be worth noting that the radio transmission frequencies for spacecraft have seldom been published, but during the early Space Age the Soviet Union announced radio frequencies for the spacecraft that performed heroic feats like photographing or hitting the Moon and putting humans into orbit. The idea was obviously that "the West" would to eavesdrop and be impressed! The United States, on the other hand, kept the frequencies for their first manned spacecraft, Mercury, rather confidential. When the two-man Gemini craft appeared, NASA started to be more open. But the Soviet Union continued to publish radio frequencies for some of their satellites and strangely enough very often included the frequencies of what must have been their most secret spacecraft – the photo-reconnaissance satellites. Admittedly, they only issued shortwave frequencies such as 19.995 MHz where quite simple and relatively undecipherable information was transmitted. But what was the purpose of regularly publishing these frequencies?

Geoff Perry and I often discussed this question, and the only explanation that we could come up with was that it was part of some 'cover story' meant to confuse the general public about the real purpose of these short-lived Kosmos satellites; i.e. the openness about their transmission frequencies was to give the impression they were scientific satellites – even if it was clear they were reconnaissance satellites. It was a matter of "deniability". If the Soviet Union publicly admitted that these were photo-reconnaissance satellites they would have been legitimate targets for a US attack by anti-satellite weapons. If the public myth that they were scientific satellites was kept up then an attack could be described as pure military aggression! But I doubt that the Soviet propaganda managers realised how much we could deduce from these beeping sounds.

LISTENING TO COSMONAUTS

Of course the radio monitoring experience that I, Geoff and the students at Kettering Grammar School desired the most was to hear cosmonauts talking to mission control (US manned spacecraft never flew over Europe in those days because the inclination of their orbits was too shallow).

Only six weeks after I picked up my first satellite signals, on 22 February 1966 the Soviets launched Kosmos 110 with two dogs on board, Veterok and Ugoljok. The transmission frequency 19.894 MHz was included in the TASS telegram announcing the launch – and it was easy to receive the signals. They were the usual FSK but with some irregular sounds at the end of the 'frame'. We have never been able to interpret these sounds, but they may have been some kind of indication of the body motion of the dogs. According to Geoff the signals sounded similar to signals from Voskhod 2, from which the first-ever spacewalk took place in March 1965. Kosmos 110 stayed in space for 22 days and this long space flight was interpreted by many as an indication that the Soviet Union was preparing for a three-week long spaceflight by a crew of two. Now, in hindsight, we know that Kosmos 110 was an unmanned version of the Voskhod spacecraft. The Soviet Union was preparing to launch Voshkod 3 during the period March–May 1966, but this never took place because resources were directed towards completing the more modern Soyuz.

Kosmos 133 was launched on 28 November 1966. Even though TASS announced the transmission frequency to be 19.995 MHz, we suspected it to be an unmanned test of a new piloted spaceship. The spacecraft had severe problems with its orientation system and when the retrorocket was fired this cut off prematurely and an automatic system destroyed the capsule by explosive charges to prevent it falling into the hands of the Americans.

The next unmanned test flight of the new spaceship took place on 7 February 1967 and lasted two days. No docking in space was planned, just a test flight to validate the design of what would become known as the Soyuz spaceship. After some difficulties in orbit, the capsule of Kosmos 140 successfully returned to Earth after two days but sank to a depth of ten metres in the Aral Sea. It was eventually retrieved and Soviet engineers somehow decided to go ahead and launch a human in the next Soyuz. By chance, Geoff found out from a friend at the University of Wales in Aberystwyth that the physics institute there had picked up unusual signals from a satellite on 20.005 MHz on four occasions while listening for signals from NASA's Explorer 22, which was a radio beacon for studies of the ionosphere, the electrically charged portion of the Earth's atmosphere at the very edge of space. The unusual signals were keyed on and off and were transmitted in 30-second bursts starting every 120 seconds. Geoff identified the source of the signals – Kosmos-140 – as it passed directly over Wales when the signals were picked up! We later discovered that TASS had specified the transmission frequency as 20.008 MHz, so the University's receiver may not have been as narrow-band as was assumed.

New spaceship but we hear nothing!

The morning news on Sunday 23 April 1967 included the fact that the Soviet Union had launched a new piloted spacecraft, Soyuz 1, at 0135 CET that same morning. On board was Vladimir Komarov. This launch was really no big surprise. I called the Swedish News Agency TT and managed to find an editor who read me the entire TASS telegram about the launch, which announced the transmission frequencies as 15.008 MHz, 18.035 MHz and 20.008MHz. The last one agreed well with that of Kosmos 140 and the observations in Wales. I spent all day at my radio of course, but heard nothing. However, Harro Zimmer, an amateur satellite tracker in Berlin, heard the cosmonaut talk on 18.035 MHz several times before noon.

Early next morning, when the spaceship approached Western Europe, I was at my radio again but heard absolutely nothing on any of the three frequencies. At Kettering the result was the same. (At that time letters were our only means of communication. Later during 1967 I started calling Geoff by telephone, but the calls went through an operator and sometimes it took hours to achieve a connection. It was very primitive compared to the situation today.)

While I was tuning around with my radio, disaster struck at 0424 CET when the Soyuz spaceship plunged to Earth with a tangled parachute and Vladimir Komarov perished. I was very disappointed and was almost discouraged from continuing my tracking, but the Space Race was entering a critical phase. Then in October a Soviet spokesman for the space programme hinted at "new victories soon". We speculated about various space spectaculars, but the real event started without much fuss. On Friday 27 October Kosmos 186 was launched. Radio Moscow's English-language service did not mention it in its evening broadcast. The evening television news on Saturday mentioned that one of the three newly launched Kosmos satellites had an orbit similar to that of Soyuz 1. I assumed that it was on the same frequency as that received in Wales from Kosmos 140, and listened to 20.005 MHz all day Sunday. A Swedish newspaper also carried a story from Britain reporting that Geoff had picked up signals similar to those from Kosmos 140. But I heard nothing.

With a desperate need for accurate information, at 1515 CET I telephoned Geoff (this was an expensive step because the call cost me about $50 US). When I reached him, Geoff said he believed Kosmos 186 may have been recovered that morning and he gave me the exact frequency of 20.008 Hz. On Monday morning when the satellite – if it was still in orbit – would have been close to Stockholm I was at the controls of my brand-new Eddystone shortwave radio. Finally, at 0917 CET I heard the typical beep-beep-signals that Geoff had whistled (!) to me over the phone the previous day. So, I now had heard, for the first time, the characteristic on-off-keyed pulse-length-modulated signals from a Soyuz spaceship, in this case flying unmanned. Was this Kosmos 186 still in space or was it a new satellite of the same type? It turned out to be both because a new satellite, Kosmos 188, had just been launched and during the morning hours the two spacecraft had docked automatically; this was an impressive feat at that time.

"Word 8" decoded

In January 1969 the Soviet Union finally conducted the docking in space that had been the objective of the ill-fated Soyuz 1 flight in 1967. Soyuz 4 and Soyuz 5 were joined and two cosmonauts (Yeliseyev and Khrunov) walked in space from Soyuz 5 over to Soyuz 4, where Vladimir Shatalov waited for them having been launched on his alone. Boris Volynov remained in Soyuz 5 and returned to Earth alone. We heard lots of signals on 15.008 and 20.008 MHz, and at Kettering a brief shortwave voice call was heard from one of the ships. It was now evident that voice signals from the Soyuz spaceships were transmitted on an entirely different frequency band and that shortwave was a seldom-used backup voice link.

But Geoff Perry made an amazing discovery that later enabled us to flabbergast journalists. He noticed a characteristic in the signals from Soyuz 5 on 15.008 MHz during its first day in space. The eighth signal pulse (known as "word 8") after the synchronization "purr" varied in length in a systematic way. Four frames in a row it was short, then at intermediate length four times and finally long four times. Then the sequence was repeated. Not until the autumn of 1969, when three piloted Soyuz spaceships were in orbit simultaneously carrying, respectively, two, three, and two cosmonauts on board, did the meaning behind the changes become evident. Geoff saw that during this mission "word 8" again varied in length. In some recordings it had two lengths and in other recordings it had three lengths.

Geoff then went back to his recordings from the Soyuz 4/5 mission and found that after the two cosmonauts transferred from Soyuz 5 to Soyuz 4 "word 8" from Soyuz 5 changed from having three lengths to a single length. In the signals from Soyuz 4 the opposite occurred. It was possible to distinguish between the spaceships because they used different frequencies (15,008 and 20,008 MHz). The conclusion was that "word 8" related to the size of the crew. In all likelihood "word 8" indicated which cosmonaut was being monitored by some other "word"; we suspected "word 4". As long as shortwave radio was used by Soyuz spaceships we could pick up the signals directly after launch, and "the naked ear" could determine the crew size by checking the length of "word 8" before the TASS launch announcement. Three lengths, three cosmonauts, two lengths, two cosmonauts: elegant, wasn't it?

Voice channels

It would be another four years before we could receive the voice transmissions by cosmonauts. During 1970 there was a single Soyuz mission, an 18-day flight with two cosmonauts. In 1971 there were two Soyuz flights that docked with the space station Salyut 1. The first docking did not work properly, but the second enabled a crew of

3.3: US satellite tracker Richard Flagg.

three cosmonauts to live on board the station for three weeks; unfortunately they perished returning to Earth when their capsule was inadvertently depressurised in the vacuum of space. We tried every known Soviet space frequency, including the VHF frequency of 143.625 MHz used by Yuri Gagarin and other Vostok cosmonauts. We heard only the on-off-keyed shortwave signals. Salyut 1 also transmitted similar signals on the same set of frequencies as the Soyuz because it was a civilian version of the military space station Almaz but using Soyuz hardware – including the radio system. But we didn't understand that then.

At the end of 1970 Geoff Perry put me in direct contact with a radio astronomer, Richard (Dick) Flagg at the University of Florida in Gainesville. Dick and his family had moved close to Cape Canaveral because he worked for Pan American Airways (the airline company) which operated the technical facilities of the military part of the rocket base. As a part of this work, Dick was sometimes temporarily assigned to Hawaii and served on ships that monitored Soviet intercontinental missiles launched from Baykonur re-entering the atmosphere over the Pacific; both Soviet and US ships were present to monitor their re-entries. His knowledge of radio technology is vast, so we started an intensive correspondence about, for example, the best radio equipment for various frequency bands. Dick helped me to acquire equipment for 922.75 MHz, which we suspected was employed by Soviet interplanetary probes, lunar probes, and piloted spaceships. This turned out to be correct, and I successfully picked up radio signals from Salyut 1 on this frequency before the station was deorbited.

SEARCHING FOR THE 'LOST DIAMOND'

When the world's first space station, Salyut 1, was launched on 19 April 1971 the Kettering Group could hear signals of the same type as those from Soyuz (on-off-keyed pulse-length-modulated carrier, CW-PDM) and on the same frequencies of 15.008 MHz and 20.008 MHz. As such, this was no surprise to us, but we did not understand why this was the case.

Nothing happens – or so we thought

Given the tragic accident when the crew of Soyuz 11 died from decompression after having spent three weeks on Salyut 1, we expected a long pause in the space station programme. After an unmanned test flight of a modified Soyuz in June 1972 there were rumours in Moscow about the imminent launch of another space station. But nothing happened – or so we thought. What really occurred was that the launch of a space station similar to Salyut 1 failed on 29 July. We had no knowledge of this at the time, nor did we know that there were two different space station projects running in parallel.

On 3 April 1973 TASS suddenly announced that Salyut 2 had been launched. We rushed to our radios but heard absolutely nothing – which was very perplexing. Then on 28 April it was announced that Salyut 2 had completed its mission and provided valuable information that could be used to design future space stations. Orbital data that we had obtained from NASA showed that the orbit of the space station had been adjusted by rocket impulses up until 9 April. We speculated that the solar panels had not deployed and that low power was the reason for not switching on the shortwave transmitters. The absence of signals on the usual frequencies was hard to explain. We grasped at any explanation. After the TASS announcement there were rumours in the trade press that there had been an explosion on board Salyut 2. It appeared as though another disaster had struck the Soviet space programme on the very eve of the launch of NASA's first space station, called Skylab.

Salyut or Soyuz?

On the evening of 11 May 1973 TASS announced the launch of Kosmos 557 with orbital elements very similar those of Salyut 1 at that point in its mission. Geoff Perry obtained orbital data from the Royal Aircraft Establishment at Farnborough showing that the TASS announcement had been issued 20 hours after the launch, which was clearly strange.

Again we rushed to our radios, and 28 hours after launch I picked up signals on 922.75 MHz with a carrier and sidebands similar to those from Salyut 1 and also an unmanned Soyuz in June 1972. These signals had not been heard from Salyut 2. But

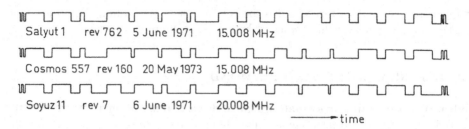

Salyut 1   rev 762   5 June 1971   15.008 MHz

Cosmos 557 rev 160   20 May 1973   15.008 MHz

Soyuz 11   rev 7   6 June 1971   20.008 MHz

time

3.4: Similarity in telemetry between Kosmos 557 and Salyut 1.

was Kosmos 557 a Salyut or a Soyuz? At that point there was no way of finding out. Furthermore, we had no information on any possible manoeuvres by Kosmos 557 to alter its orbit. But on 20 May there was a breakthrough. Jan-Ola Dahlberg in Malmö, in southernmost Sweden – the only member of the Kettering Group that I recruited – heard signals on 15.008 MHz. But this could not distinguish between a Salyut or a Soyuz because both used this frequency! We heard the signals on 15.008 MHz for two days until the spacecraft burned up in the atmosphere over Australia. Obviously this was another embarrassing failure. But Geoff analysed the signals and decided it was the same type of vehicle as Salyut 1 despite an assertion by the Soviet Academy of Sciences that Kosmos 557 had nothing to do with the piloted space programme – a statement that was then already seen as an outright lie.

Probably it was too embarrassing to admit that yet another space station had failed – at a time when Skylab had just been launched. The designation Kosmos 557 was an obvious cover name and the long interval between the launch and the announcement showed that something had gone wrong and that the designation was an attempt to hide the true purpose of the mission.

So Kosmos 557 was a space station and it transmitted the same kind of signals as Salyut 1, but what was Salyut 2? We had to wait a year to find out.

Military space stations

At 2338 CET on 24 June 1974 Salyut 3 was launched from Baykonur. Again, just as with Salyut 2, we heard nothing on the known frequencies. We feared that we would again be sidelined during a major Soviet space adventure! In the evening of 27 June I talked to Geoff on the telephone and we analysed a small item in an issue of *Aviation Week* from the previous September. I suggested this should be interpreted to indicate that the signals from Salyut 2 resembled the frequency-shift-keyed and pulse-length modulated (FSK-PDM) shortwave signals that we were picking up from the photo-reconnaissance satellites in the Kosmos series.

Geoff proposed checking the known frequencies near 20 MHz. I volunteered, and set the alarm clock to 0145 CET in order to catch the pass of Salyut 3 that would be nearest Stockholm. The frequency range 19-20 MHz was almost completely 'dead' at that time of day; no far-off radio stations could be heard. This meant that if Salyut 3 transmitted in this band it would be heard only when above my horizon and within line-of-sight. This would certainly facilitate identification of any signals received. As I tuned slowly upwards from 19 MHz at 0207.45 CET, I got the typical FSK-PDM signals that we were used to hearing from the photo-reconnaissance satellites. The signal was very strong, and faded into the noise at 0212.50 CET. I measured it to be 19.944 MHz. Both the frequency and the signal type were distinctly different from those of Salyut 1.

Were there two types of Soviet space station sharing the same Salyut programme name? It was an unavoidable conclusion based on the signals I had just picked up! If it were true – how could the Soviet Union afford two expensive programs? And what purposes did they serve? Had the Soviets really planned to have two space stations in orbit simultaneously in April-May 1973 in order to upstage the launch of Skylab?

In December 1974 Salyut 4 was launched and it transmitted signals like those from Salyut 1 and Kosmos 557. Furthermore, it was in a higher orbit than Salyut 3. I wrote to the British Interplanetary Society's *Spaceflight* magazine suggesting that the Soviet Union had a "military" space station programme using low orbits and a civilian space station programme in a higher orbit. I revealed that the two types transmitted different kinds of radio signals, and also pointed out that these two types could not possibly be equipped with the same set of subsystems because the failure in orbit of Salyut 2 did not seem to affect the launch shortly thereafter of Kosmos 557.

By using radio observations and common sense we had cracked the 'cover name' of Salyut and exposed a really big secret – the two space station programmes that we later learned were called DOS in the case of the civilian programme and OPS Almaz (Diamond) for the military one. So, in June 1974 we found "the lost diamond" – the military space station. Once a crew was launched to Salyut 3 we found more evidence of its military purpose.

"The Golden Eagle" calls

When we had succeeded in finding Salyut 3 on shortwave, we turned our attention to finding out when a crew would be launched to the space station. By using the orbital data supplied by NASA and the Royal Aircraft Establishment, together with our radio observations, we could determine its ground track on a map of the Earth. It flew over every area of the globe between 51.6°N and 51.6°S every day, but the track repeated exactly every four days. Days when the ground track passed closest to the rocket base at Baykonur in Kazakhstan presented opportunities for a launch to the station. On 29 June media reports said that a Soyuz would be launched on 1 July, but our analysis showed that the correct date was 3 July. We also computed the launch time as 1953 CET, and we expected to receive signals from the new spacecraft on 20.008 MHz at 2002 CET.

My wife Inger and I had recently moved to a suburb of Stockholm, but I hadn't installed any permanent antennas yet. The antenna for the voice channel on 121.75 MHz was attached to a broomstick that was clamped to the balcony barrier and the antenna for 922.75 MHz was on a camera tripod on the balcony! I must admit that I got goose bumps when I heard the weak, but clear, beeps from a Soyuz at 2001.50 CET! I immediately telephoned Geoff, and he informed the media in the UK that a

Soyuz had been launched at 1953. About an hour later I heard the signals again and they were much stronger. "Word 8" varied between two lengths, indicating a crew of two cosmonauts!

At 2254 CET, three hours after launch, I picked up my first clear voice signals from a Soviet spaceship. The call *"Zarja, ya Berkut, pryom"* (*"Dawn, this is Golden Eagle, over"*) was strong in the loudspeaker of my homemade receiver. So the commander had the call-sign *"Berkut"*. I thought I remembered it, and thumbed through my old newspaper clippings. Yes! Pavel Popovich had used it during his Vostok 4 flight in August 1962. The Soviet media still had not said a word about the launch, but the Kettering Group could tell the world's press that Pavel Popovich and an as yet unidentified colleague were launched at 19:53 CET! A rather neat signals intelligence operation.

At 0035 CET the Reuters news agency sent a telegram from Moscow in which it reported that they had called the Soviet information ministry and asked if a piloted spaceship had been launched. The official who answered the phone call had replied, *"Yes, they have put one up"*. I continued hearing *"Berkut"* talk to Earth on two more passes over Europe and it took until 0500 CET for TASS to issue the official launch announcement. On board the spaceship Soyuz 14 was indeed Pavel Popovich and the new cosmonaut Yuri Artyukhin.

More mysteries turn up

The following morning Dick Flagg in Florida picked up voices from Soyuz 14. The Soviets had evidently placed a ship in the Caribbean to communicate with the space station crew. Dick heard the spacecraft call *"Komarov, ya Berkut"* (*"Komarov, this is Golden Eagle"*). The Soyuz was obviously talking to *Cosmonaut Vladimir Komarov*, one of the gigantic floating antenna ships supporting the mission. Later Dick would find that the tracking ship *Cosmonaut Yuri Gagarin* had anchored off Nova Scotia in order to communicate with the crew in space.

Soyuz 14 docked with Salyut 3 at 2300 CET on 4 July. It took some time for the crew to transfer to the space station, but after hearing them communicate on 121.75 MHz at 0312 CET on 5 July all signals from Soyuz 14 ceased. After moving into the space station they switched to another voice channel, and we were once again cut off from the space adventure!

We decided to try three known Soviet radio channels: 142.4 MHz intended for the Apollo-Soyuz Test Project in the summer of 1975, the old lunar probe frequency of 183.6 MHz, and Gagarin's Vostok channel on 143.625 MHz. I checked 183.6 MHz and heard nothing, but Jan-Ola in Malmö tried 143.625 MHz and found out that this was the correct guess. Even though Jan-Ola's amplitude modulation receiver badly distorted the frequency modulated (FM) signals from the space station it was clear that we had found the voice channel again.

In Florida Dick Flagg quickly built an FM-detector for his receivers and started to report what the cosmonauts were talking about courtesy of a professor in the Russian language at the University of Florida. They were talking about *"film cassettes"*. Was that film from a reconnaissance camera?

"Word 7" goes long

Two weeks after the docking, I began to suspect that the flight was about to end so I started to listen on the shortwave frequency of 20.008 MHz. At lunchtime on 19 July 1974 I heard the typical signals and noticed that "word 7" was long. We knew from previous Soyuz flights that this meant a return to Earth was immediately imminent. The shortwave transmitter was turned off at 1250.45 CET. Another rule-of-thumb of the Kettering Group was that the actual touchdown would occur 30 minutes after the shortwave signal switched off. So at 1301 CET I called the Swedish News Agency TT and told them that I thought the landing would occur at 1320 CET. Later TASS announced that the crew had landed in Kazakhstan at 1321.36 CET. So the rule-of-thumb was quite accurate!

TRACKING MOON MISSIONS

By chance 1972 was unusually productive for my satellite monitoring hobby. This was particularly true for receiving signals from spacecraft exploring the Moon. In February of that year I succeeded in hearing signals from a Soviet space probe that was returning to Earth with a sample of lunar soil.

The final Luna flights

The International Telecommunications Union reserved the frequency band 183-184 MHz for use in space, but as far as is known only Soviet lunar and planetary probes have used it. The first probe to fly close to the Moon, Luna 1, the first space vehicle to ever hit the Moon, Luna 2, and the first to photograph the hidden side of the Moon, Luna 3, all of which were launched in 1959, transmitted on 183.6 MHz. Their signals were picked up by, for example, the giant radio-telescope dish at Jodrell Bank near Manchester in the UK. The frequency was also used during the first Soviet attempts to soft-land on the Moon in 1965 and 1966. The first spacecraft to land on the Moon in February 1966, Luna 9, transmitted conventional 'wire photo' images of the lunar surface on 183.538 MHz.

The third, and final, generation of Soviet lunar probes appeared in the autumn of 1970 when the Soviets launched Luna 16. It landed on the Moon and, under remote control from Earth, sampled the soil and launched a capsule to deliver the sample to Earth. To our surprise, the Soviet news media announced that the return vehicle used the frequency of 183.6 MHz. It appeared that the transmitter was used in a system for accurately determining the return trajectory in order to calculate precisely where the 60-cm-diameter capsule carrying the lunar material would parachute to Earth. The TASS communiqué was correct, since Jodrell Bank could follow the first unmanned return flight from the Moon on 183.538 MHz up until 12 minutes before the capsule touched down in Kazakhstan.

I had bought a frequency converter for 183.6 MHz in 1969 from my "purveyor", Vanguard Electronic Laboratories. The converter changed the 183-184 MHz band to 0.6-1.6 MHz which I intended to tune using my old HE-30 receiver.

Patience needed after Luna 16

When Luna 16 was launched I was unfortunately serving with the military in the far north of Sweden and could not get home in time for the return flight to Earth. I had to be patient! Afterwards I started to analyse the possibility of picking up signals from another mission of this type. I have found a typed note dated 1 December 1970 in which I drew the conclusion that it would be easy to pick up signals during the last day of the return flight. The second successful unmanned return flight from the Moon was performed in February 1972 by Luna 20. The spacecraft weighed 1,880 kg when it landed on the Moon on 21 February. After taking samples of the lunar soil a return vehicle lifted off at 2358 CET on 22 February. By assuming that it would follow the same trajectory as its predecessor it was easy to compute when it would arrive.

Three days later, on 25 February, Luna 20 approached the Earth, and from 0600 CET I was at my radios in my apartment on the tenth floor of a high-rise apartment block north of Stockholm. I had set up an antenna on the balcony. It was actually a TV antenna for the old channels 7-11. As I began to tune looking for a signal near 183.6 MHz, I attempted to steer the antenna to the place in the sky where I thought the space probe would be. It would take until 1740 CET before I picked up a carrier on 183.54 MHz that I suspected came from the spacecraft. The signal sounded like a steady tone that was interrupted every 20 or 40 seconds by another tone, mostly eight seconds long, on a frequency offset by 4 kHz. The change in frequency was gradual and took about 1.3 seconds. What did these changes in frequency signify? I didn't understand that in 1972 but I recorded the signals and saved them.

As Luna 20 approached Earth the signal strength increased, becoming very strong, but disappeared suddenly when the landing capsule and the return vehicle (which had separated at an altitude of 50,000 kilometres) set below the horizon in the southeast at 2000.10 CET. The capsule landed in Kazakhstan by parachute in a blizzard at 2012 CET.

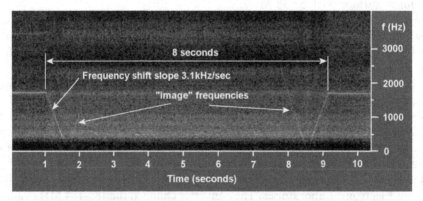

3.5: A spectrogram of Luna 20 from 25 February 1972.

Mystery finally solved after 30 years

Thirty years later I transferred the tape recording of Luna 20 to a digital sound file and analysed it with audio editing software that could show spectra of audio signals on a computer screen. I was really surprised to see the image. The frequency changes follow a straight line! When I saw the spectrogram I recalled having read a German book about space radio systems in the library of my employer – the Swedish Space Corporation. In this book was a system for measuring the distance to a space vehicle by up-linking a radio signal to it that varies in frequency and having it send the same signal back to Earth. By measuring the phase difference between the uplink and the downlink it was possible to calculate the distance to the spacecraft. That must have been the purpose of the strange signals that I heard. In a book featuring interviews with veterans of the Soviet space programme, Leonid Gusev described how he had designed such a system for probes like Luna 16. Bingo! But it took thirty years to understand what I had heard from Luna 20.

A CALL FROM SWEDISH INTELLIGENCE

Electronic intelligence is an inevitable description of my space tracking hobby, but I did it in order to understand current space events and their context. Naturally, various agencies doing similar work professionally wondered what I was up to.

Space magazines marked 'secret'

One day in the spring of 1974, I got at telephone call at work from Swedish military intelligence. The person calling was Lt Col Henric Victorin. He is a really fascinating person, and a veteran of Swedish aviation intelligence-gathering who flew planes on daring reconnaissance missions. He was also equipped with a disarming self-irony towards all things military. Henric's task was to try to understand the wider purpose of the Soviet Union's military space activities, and I suppose there were very few of his colleagues who could do that. I visited his office several times. It was situated in a very posh apartment in Stockholm's upper-class district that served as a sort of 'cover address'.

On his bookshelves I saw binders with my own *Spaceflight* articles about satellite tracking. That in itself was no surprise, but to find them all stamped 'SECRET' was! To avoid the problem of keeping straight in his mind what he had learned from secret sources and from open sources, he found it much easier to regard everything – even material in the public domain – as secret. This may sound absurd for someone outside the secret world, but it is actually a rather logical approach. Anyway, Henric was the master of the "calibrated indiscretion". What did we talk about? Let me just say we helped each other!

'Donald Duck' in space

In the summer of 1976 the second space station of the military Almaz type, Salyut 5, was in orbit. Just as during the cosmonaut visits to Salyut 3 two years previously we easily picked up frequency modulated voice from the crew on 143.625 MHz. It was fun to hear small segments from their daily life on the station. With a dictionary I tried to interpret words and phrases. I observed 35 passes of Salyut 5 with its voice transmitter switched on. Out of these passes, the transmissions from twelve were impossible to interpret. By looking at the spectrum of the voice signal on the little spectrum analyser connected to the receiver it was apparent that the signal spectrum from the space station now had two distinct peaks and the loudspeaker only emitted noise. The voice channel was obviously now encrypted! This was another indication of the military nature of the Almaz space stations.

The voices from the civilian space stations were also distorted in order to 'hide' what was said. When Dick Flagg passed through Sweden on a holiday trip in 1978 we together heard a cosmonaut on the space station Salyut 6 use voice scrambling on the 121.75 MHz radio link that made his voice sound like Donald Duck, rendering it unintelligible. We agreed that this was generated by inverting the voice spectrum; i.e. transmitting the low notes as high notes and vice versa. Dick (an electronic wizard) started asking me what kind of components there were in the electronics lab of the Swedish Space Corporation. Was there an analogue multiplier? Did we have any good filters for cutting off unwanted frequencies? After some thought I reckoned we had everything he asked for. We jumped into my car and headed there!

In just a few hours' work in the lab we assembled, tested, and used an electronic circuit that removed the voice scrambling. In the descrambled voice signal we could hear that the cosmonauts were speaking Russian but it was still difficult to hear what they said. It would take twenty-five years until a Russian-speaking Internet contact was able to make out what the crew said. At 1705 CET on 9 July 1978 one of the cosmonauts said: "I have checked the cargo ship as well. On the cargo ship there is no... There is just the same Eunoctin." Eunoctin (Nitrazepam) is a drug used in the treatment of moderate to severe insomnia. Soviet space officials later admitted that this voice scrambling was to keep discussions between the cosmonauts and doctors on the ground private – not only from eavesdroppers like myself but also from staff in the control centre in Moscow and other tracking stations.

In the middle of the 1970s another enthusiastic and very young man from Texas called Mark Severance joined our group. When there was a Soviet tracking ship off Cuba he could hear cosmonauts talk on VHF and he made some really high-quality recordings of the crew of Salyut 5. After his college graduation he eventually ended up at NASA's mission control in Houston. When the International Space Station (ISS) started to be assembled in orbit, Mark served in the mission control centre in Moscow as a NASA liaison officer. The Kettering Group finally had a person inside the 'holiest of holies' of Russian spaceflight!

Chasing frequencies with a tape measure

During Dick Flagg's visit to Sweden in the summer of 1978 there was a Soviet space exhibition in Stockholm. We decided to pay it a visit and check out all the antennas on the full-scale models of Soviet spacecraft in the exhibition. The remarkable thing about these models was that they appeared to be entirely correct in all dimensions – it seemed that they were manufactured using the original drawings. We walked from spacecraft to spacecraft and measured the sizes of the antennas that we could reach in order to deduce their operating frequencies. As we were doing this, we saw that the Soviet staff were trying to take pictures of us when they thought we wouldn't notice. So one of us watched the watchers as the other measured antennas. Perhaps this kept our faces out of a KGB photo archive!

Super-secret satellites

We were convinced that the hunt for new radio frequencies would yield new insights into the epic battle for supremacy in space between the superpowers. In addition to using tape measures and Swiss radio spectrographs, Dick and I agreed that he would help me acquire receiving equipment that would make it possible for me to monitor a wide frequency range. At a surplus dealer he was able to get hold of a radio receiver that had been used at the Kennedy Space Center and covered the range 55-260 MHz. Furthermore, it was equipped with a so-called panoramic adapter, a small screen on which the signal strength could be monitored as a function of frequency (i.e. near the tuned frequency of the receiver). Dick also modified the radio so that it could receive frequency-modulated signals such voice transmissions from Soyuz spaceships. This receiver (a NEMS-Clarke 2501 A) was a real "monster" – being equipped with good old-fashioned vacuum tubes it was big and heavy. It arrived by boat in the spring of 1975 and I soon got it up and running. When I used it for the first time it was just like stepping into a whole new world.

A very peculiar type of signal that I discovered rather quickly, which was emitted by several Soviet spacecraft, appeared in the loudspeaker as a high-pitched whizzing sound. On the panoramic adapter screen it looked like a diffuse 'noise hump'. After much contemplation we guessed that it was some kind of wideband pulse-modulated transmission. Classical signal theory says that if the signal is very wide in frequency, then the emitted pulses are very short. But to be able to receive this signal correctly and demodulate it, I required a receiver with a much wider bandwidth. At the end of the 1970s I acquired yet another receiver in the NEMS-Clarke range (Model 1302 A) that Dick had bought as surplus equipment. When I connected an oscilloscope to the receiver the short pulses immediately appeared on the oscilloscope screen. Every 80 microseconds a pulse was transmitted. Within this interval another pulse was issued. The position of this pulse within the interval represented a measurement. If the pulse was transmitted in the middle of the 80 microsecond interval it meant 50 per cent of full scale. If the pulse was transmitted right at the end of the interval it meant 100 per cent. On the oscilloscope picture we could see measurement channels. The left was at 50 per cent of full scale and the

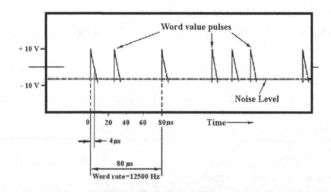

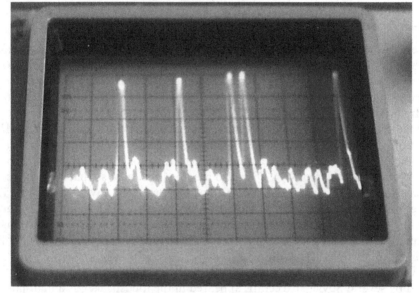

3.6: PPM-AM signals from Kosmos 1286.

right was at about 10 per cent of full scale. But only the Soviet engineers knew which parameter each 80-microsecond interval actually represented.

This type of data transmission by radio – known as telemetry – had been used by German engineers in V2 missile tests. After the war, it was used for some time at the US rocket base at White Sands, New Mexico, for V2 and Aerobee research rockets. Soviet engineers refined the method and it was widespread in their space programme for many years. This is called Pulse-Position Modulation – Amplitude Modulation (PPM-AM). The designation AM means that the pulses were transmitted as varying amplitude (signal strength) rather than as varying frequency.

THE CONTROL SPUTNIKS

Kosmos 1286 was a satellite in an orbit at an altitude of about 435 kilometres and its purpose was to monitor radar emissions by US naval forces. After the Cold War we discovered that the Soviet designation of these satellites was US-P (Upravlenniye Sputnik Passivny – 'Control Sputnik Passive') and they were part of a larger system designed to keep track of US aircraft carriers for targeting Soviet cruise missiles. In case of an armed conflict, the idea was for the Soviets to knock out the main elements of the US fleet. This system also involved satellites equipped with radar for imaging large areas of the ocean to show the position, heading and size of these naval vessels. These were called US-A (Upravlenniye Sputnik Aktivny – 'Control Sputnik Active') and were powered by a nuclear reactor carrying 30 kg of uranium fuel. For the radar to be effective the satellites had to fly as low as 250 kilometres, where atmospheric drag caused rapid orbital decay. The satellites were regularly boosted back up to 250 kilometres by small rocket impulses, but when the propellant ran out the radioactive fuel left was normally separated and boosted into a safe 1,000-kilometre orbit which would be stable for more than a millennium, sufficient to allow the radioactivity to decay to reasonable levels.

3.7: Geoff Perry tracking a Soviet satellite.

We could monitor both US-P and US-A. I easily picked up PPM-AM on 166.0 MHz from both types, and most members in the Kettering Group could also hear signals from US-A on 19.542 MHz, which was actually a frequency announced in Soviet media!

Two failures – one very embarrassing

The US-A satellites worked in pairs, one trailing the other along the same orbit with a time difference of 25-40 minutes. The idea was that two consecutive images of an area of the ocean would show in which direction ships were moving. In September 1977 the satellite pair Kosmos 952 and 954 were launched. At some time around 1 November 1977, Kosmos 954 lost the ability to maintain its operational altitude by small rocket impulses. Its altitude started to decrease gradually, but no immediate boost of the reactor to the higher orbit occurred. US authorities were soon aware of what was going on. On 6 January 1978, the US Space Command reported that radar observations of the satellite showed it to be tumbling. Now it was impossible to boost the reactor to the higher orbit. It was destined to fall to Earth within a few weeks. On 14 January 1978 the Soviets confirmed that the satellite had "lost pressurisation" on 6 January and that its reactor had 50 kg of uranium-235. On 18 January they explained that the reactor would not "go critical", and that it would burn up during re-entry. On 23 January Soviet authorities announced that the

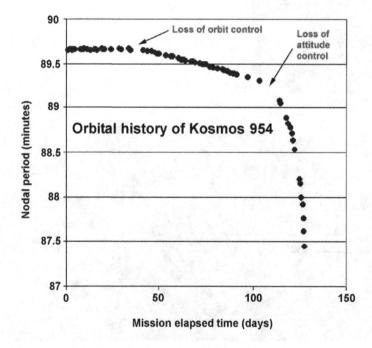

3.8: Kosmos 954's slow decay from orbit.

satellite would return to Earth the next day. Debris from Kosmos 954 fell on the Great Slave Lake in Canada at 1153 CET on 24 January.

An enormous search-and-rescue effort was started to find any radioactive debris. The search covered an area of 600 sq. km between the Great Slave Lake and Baker Lake in Canada's Northwest Territories. Twelve pieces of the satellite were found – the largest of which weighed 18 kg and could deliver absorbed radiation doses of up to 1 sievert(Sv)/hour. (The background radiation contributes several *milli*sieverts per year! The 'safe' dose limit for radiological workers was 50mSv/year.) The Canadian government sent the Soviet Union a bill for six million dollars to cover the cost of the cleanup, and eventually the Soviet Union paid half of this. It was a very public failure of a Soviet spacecraft. But after a pause of two years the US-A satellites started flying again.

History repeats itself

On 30 August 1982 Kosmos 1402 was launched while Kosmos 1365 was operating in the low orbit. When Kosmos 1365 was decommissioned and boosted to the high orbit on 27 September, we reckoned that another satellite launch was imminent. On 2 October Kosmos 1412 was launched but Soviet engineers had difficulties with it and had to boost its reactor to the higher orbit in November. Kosmos 1402 flew on alone, but at the end of December dramatic events occurred.

Both Geoff Perry and I kept a wary eye on the US-A satellites by collecting their orbital elements from the Xeroxed lists of orbital elements that NASA mailed for free to interested parties! They were mailed from a NASA centre several times a week and took five days to reach us. So we watched over these satellites with 5-6 days' delay. I cut out the little strips with the orbital elements of a satellite at a particular moment. Then I glued these strips on index cards. I still have thousands of such index cards in my files. One of Geoff's pupils, John Corvesor, also collected orbital elements for Kosmos 1402.

On 3 January 1983 an American journalist telephoned Geoff Perry and asked if he was aware of any difficulties with Kosmos 1402. Geoff could only reply that "on 26 December everything was in order". While we were waiting for the next envelope of orbital elements, Geoff called Aston University on 4 January. They received orbital data from NASA via telex and confirmed to Geoff that Kosmos 1402 had split into three parts, but provided no orbital elements. The following morning (5 January) he called me and asked if I had received a new envelope from NASA and whether I had noticed the new objects in orbit. I had indeed, and had already glued the strips for the objects I was watching onto their index cards. But in the waste-paper basket I found orbital data for all three fragments of Kosmos 1402; the main object A and the two new objects B and C.

The orbital data strips showed that the new objects had appeared in the morning of 28 December and were still in the lower orbit. A new international incident was under way! Jan-Ola Dahlberg in Malmö had heard signals from the satellite on 19.542 MHz on 30 December (and once more the following day) and my calculations showed that the source of the signals was in the low orbit. We knew from monitoring the previous satellites of this type that the shortwave transmitter was in the part of the satellite that was boosted to the high orbit. So, the Soviet engineers had failed to boost the reactor to the higher orbit. Geoff could now issue a message to the press about the crisis for Kosmos 1402.

Another Kosmos disaster?

I immediately provided this news into Swedish Defence Headquarters. The question now was: Was this going to be another environmental disaster like the fall of Kosmos 954?

In Radio Sweden's news at 1230 on 6 January Soviet media were reported to have denied that there were any problems with Kosmos 1402, which continued "on a pre-planned course and presented no danger", but the following evening they did admit that the satellite was coming down:

> "... On Dec 28 the satellite's active mission ended and in accordance with the flight program it separated into individual fragments upon command from the ground. The purpose of this was to isolate the active part of the reactor so that it would burn in the dense layers of the atmosphere, the radioactivity levels following this remain within the limits of the natural background..."

Soon the brown envelopes from NASA started coming again after the Christmas and New Year perturbations in the postal service. Using data from these envelopes, another member of the Kettering Group, Pierre Nierinck in Dunkirk, computed that the density of the three objects could be described by the following scheme: B:A:C = 1:10:25, which made the densest objects A and C. Pierre estimated that the A object would re-enter the atmosphere on 23 January plus or minus 4 days and that C would do the same on 7 February plus or minus 7 days. This agreed well with computations made by Geoff.

Group of experts

After a short business trip I was back in Stockholm on 13 January and that morning I got a call from a fellow student from the Royal Institute of Technology in the 1960s, Lars-Erik de Geer, now at the Swedish Radiation Safety Authority. He wanted me to participate in an expert group on Kosmos 1402. (Lars-Erik had proved that U-137, a Soviet 'Whiskey'-class submarine that ran aground in southern Sweden in 1981, was carrying nuclear-tipped torpedoes by rowing a boat around the vessel with radiation detectors!) Shortly after Lars-Erik's call, I received a formal request from Defence Headquarters and by 1500 I was at a meeting about Kosmos 1402. My task was to back up military intelligence expert Werner von Francken.

Before noon on Saturday 15 January I called Geoff Perry and argued that the A object lost altitude faster than if the heavy reactor was still attached (the denser an object the less it is affected by atmospheric drag). I argued we were concentrating on the wrong object, and if we wanted to know where the reactor was going we should concentrate on the C object. Geoff was sceptical and said that there is always some extra piece that is dropped off in the higher orbit. At 1800 a journalist called Geoff and informed him of a TASS announcement that the reactor core would burn up in the atmosphere in the middle of February! So, it was the C object as I had proposed!

I called von Francken in the evening and told him this crucial news, but he did not seem interested. I was so upset that I called a high officer in Defence Headquarters, because I didn't want them find out about this important information by reading it in the newspaper the following day.

In the TASS message it was emphasised that the reactor core would be destroyed during re-entry and the radioactivity would be distributed in the atmosphere to such a degree that it would not be detectable against the natural background radiation. This sounded fine, but there could be contaminated material in the A object. We continued to refine the date of re-entry.

Soviet engineers had redesigned the satellite so that its dangerous nuclear material would be ejected from the reactor if the ascent to the high orbit failed. This confusion around Kosmos 1402 showed that Western intelligence had not quite come to terms with this redesign, despite the appearance of the extra object (the reactor core) in the high orbit. This redesign showed that Soviet engineers knew that the previous design could never be made to fail "safely".

An English janitor helps

During the end of the work week 17-21 January, I was in contact with von Francken again and we agreed that the A object would return to Earth on 23 January. But the important aspect for us was the time of day that the object re-entered the atmosphere – would the last ground track pass over Sweden or not? We needed to process data during the weekend in order to determine this. At lunchtime on Friday 21 January, Geoff called with the orbital data that would be the last that we could obtain before the decay of the A object.

While all authorities took the weekend off, the Kettering Group geared up. I called everyone in America that I knew and fished for orbital data, without result. But Max White, a young member of the Kettering Group, was visiting Aston University on his own business and managed to persuade a biology student to enter the Satellite Unit's office to check what was on their telex machines. This student convinced a janitor to unlock the telex room, identified the correct machine and read the orbital data for the A object to Max via telephone! I got the numbers from Geoff and gave them to von Francken, who used them to prepare the emergency order to the county governments in Sweden. The forecast at that point was that the A object would decay at 2125 CET on Sunday 23 January. On that day I got more data from Geoff and we finally agreed that the decay would occur at 2330-0330 CET. I informed Defence Headquarters via telephone. At 0020 CET Geoff called and told me that the A object had decayed over the Indian ocean at 2322 CET. The satellite had missed coming down over Sweden by a mere twenty minutes.

On 7 February the reactor core decayed. But was all this commotion necessary? Hadn't the redesign of the US-A made it a much less dangerous system than the old version used in the Kosmos 954 mission? The simple truth was that no one knew for sure and the Soviets had not been crystal clear in their description of the redesign. But it was great fun to use the capability for quick reaction work of the Kettering Group in a matter of national importance and I certainly felt proud of our efforts.

CONCLUSIONS

Why was it so addictive to track satellites launched by a secretive society? I think the attraction was that we had a sense of participating in great events whilst applying the scientific method to deduce the characteristics of the satellites we were tracking and then predict the future twists and turns in the Soviet (nowadays a much less secretive Russian) space programme.

Geoff was very good at making it fun to use deductive logic and basic physics and maths. I also think that Geoff and Derek took a certain pride (which I share) in doing all these things with relatively simple means. Our radios and computing method were almost always at "the trailing edge of technology", but I think Geoff regarded this as yet another advantage. The use of simple methods places emphasis on the deductive and "basic physics" aspects of the task. The few radios in the corner of the lab really epitomised what I think should be the credo of every scientist or engineer: "Do more with less".

4

Cosmonauts who weren't there

by James Oberg

In the years between the end of the Apollo programme and the first orbital flights of the space shuttle, when I was on the Mission Control team in Houston preparing for the first launch of Columbia, one of my additional duties was to provide background briefings for new hires. I found that a particular set of "space history" slides made one audience especially nervous. It wasn't what the slides showed but what they did not show.

The pictures were of groups of Russian cosmonauts smiling confidently for the cameras but what made the audience laugh, at first, was that subsequent versions of the very same group photographs had gaps. Faces clearly seen in the first versions had vanished to the retoucher's airbrush. My most nervous audience was the new 'space shuttle astronaut selection', 35 men and women chosen in 1978 to supplement the two dozen Apollo veterans and as-yet-unflown rookies from that era. They were sobered to realise the apparent implication of the forged Russian cosmonaut pictures – if a space trainee screwed up, he (or she) could just disappear. To prevent that from ever happening to themselves, they all vowed not to screw up. As an added defence against erasure, they joked that in any group photo sessions they would entwine their arms very tightly with each other.

And it really was funny. Here were clumsy Soviet propagandists obviously trying to conceal the existence of several individuals who had been members of their early cosmonaut teams. But since both versions of the photographs – 'before' and 'after', and in some cases several 'after' versions – were published in different books, the frauds were all too readily spotted. All that they had succeeded in doing was to raise the level of interest in who it was they desired to hide – and why.

THE SEARCH BEGINS

With the launching of Yuri Gagarin on 12 April 1961, and the subsequent flights of Titov, Nikolayev, Popovich and other space pilots, it became evident that there was a cadre of Soviet cosmonaut trainees – the equivalent of NASA's 'Mercury Seven'. Its members were never named in advance, and even the size of the group was not revealed. With nothing to go on, observers were left to offer speculations, rumours, and bad guesses based on the names of test pilots and space equipment testers who were occasionally named in Moscow newspapers.

Even the external configuration of Gagarin's Vostok spacecraft and its booster rocket remained shrouded in secrecy until the fifth anniversary of that flight. Further images of the hardware (and of cosmonaut training) were released for the Soviet celebrations in 1967 for the fiftieth anniversary of the Bolshevik coup that created the USSR. Some unknown faces were noticed in the training films but little more could be gleaned from them until the tenth anniversary rolled around.

What was hidden from the world in those years was the existence, and fate, of other would-be cosmonauts. Men with secret names such as Bondarenko, Nelyubov, and Anikeyev. But as the proverb goes, honesty is the best policy because in the long run nobody has a good enough memory to be a successful liar. We were to learn that this earthly rule applied to outer space as well.

4.1: James Oberg (right) and Charles Vick search through Soviet space photos in 1973.

First cracks in censorship

In 1971, a remarkably candid and well-illustrated book by the prominent Russian journalist Evgeny Riabchikov appeared entitled *Russians in Space*. The US edition, published by Doubleday in NYC, contained the clearest photographs of Gagarin and his teammates in training that I had ever seen. The following year, while I was on the faculty of the Department of Defense Computer Institute (DODCI) in Washington DC, I used my spare time to search through Soviet movies, books and magazines for photographs of cosmonauts. I had access to the periodicals files of the Library of Congress and to film vaults of several national news archives, including Bara Studios, which had 16-mm copies of several hours' of Soviet space propaganda newsreels. As a result, I assembled a growing collection of 'faces' of apparent cosmonaut trainees who had not yet flown.

I published these images in the British Interplanetary Society's *Spaceflight* magazine in 1973 with the provisional designations "X-1", "X-2" and so forth. The process was laborious and the payoff – identifiable photographs of cosmonauts who had yet to fly – was of interest only to specialists and spaceflight nuts such as myself. By then we had received indications that twenty men had been in the first group but, as of 1969, only twelve of them (plus members of later groups) had flown into space. So who of the unflown men might be next? The answer came like the proverbial bolt from the blue and it was – "none of them". This dramatic moment, the most exciting event so far in the lengthy investigation, came in November 1973. I still remember it vividly.

While lecturing to a computer management class at the navy base in Norfolk, Virginia, I visited the home of colleague Charles P. Vick. He proudly showed off some new acquisitions in his library of Soviet space books and I browsed through their photo sections looking for new 'unflowns'. In the middle of a new book called *Shagi V Kosmos* (*Steps into Space*) I stopped, puzzled. A group photograph of some cosmonauts (taken just after Gagarin's mission in 1961) looked 'wrong'. I recognised it as the same scene that had been shown in Riabchikov's book two years earlier. The same scene, but not quite. Vick had that book also and quickly pulled it from the shelf. Within seconds we'd found the image and laid the two books side by side. We whooped with startled glee.

Cosmonaut 'X-2' vanishes!

There was no doubt about it – a man had been removed. The figure in the back row had been replaced by some photo retoucher's guess as to what would have been seen behind him. He was designated 'X-2' in my original roster of 'unflown cosmonauts' and would now probably never be flown. The man had clearly been a close associate of Yuri Gagarin, first man in orbit in 1961. In images I had already collected, 'X-2' sat next to Gagarin when the official review board selected him for the flight and then rode with him in the bus to the launch pad. In grainy 16-mm newsreels he attended the same training sessions as Gagarin and colleagues who made later orbital flights. It was obvious that he was a cosmonaut, but 'X-2' never made his

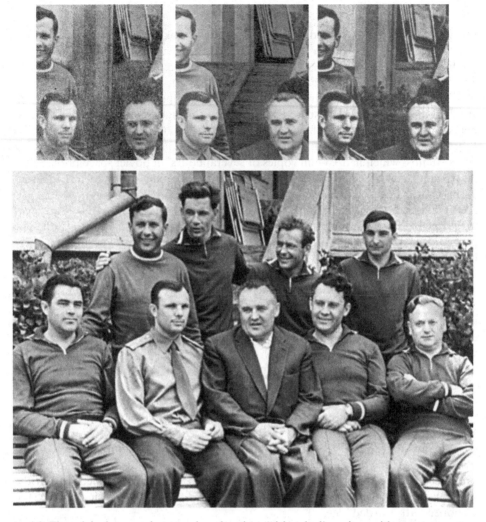

4.2: The original group photograph and various 'airbrushed' versions with a cosmonaut removed.

own space mission. In terminology invented for earlier Soviet photo forgeries, he was an "un-person". He had been 'liquidated' – maybe even evaporated. The 'X-cosmonaut' had become an ex-cosmonaut.

Ultimately, of course, this blatant cover-up backfired. Fortunately, the side-by-side publication of these 'before and after' views in Western space magazines in the 1970s made the clumsy fraud the laughing stock of space experts outside the Soviet Union. The sound of that laughter was clearly evident inside the USSR, as a series of official 'space history' books tip-toed around the question of 'unknown cosmonauts'.

Phony explanations

The Soviets later grudgingly produced an "explanation" for these extra cosmonaut names and faces. A 1977 book by pioneer cosmonaut Georgi Shonin was the first to disclose the existence of eight "dropouts" from the first cosmonaut class of 1960. It revealed only their first names – Ivan, Dmitri, Grigori, Anatoli, 'Mars', and a trio of Valentins. Shonin's book (and several later books by cosmonauts) provided sketchy accounts of their departures, which purportedly were due to medical, academic and disciplinary reasons – clearly indicating that all eight had left the programme alive. Shonin even provided a two-page character sketch of "young Valentin" (the horribly doomed Bondarenko, we later learned) without any hint of tragedy.

In another official non-response, cosmonaut Alexei Leonov was shown the picture of a 'missing cosmonaut' (Nelyubov, it turned out) by a Dutch journalist and gave a phony explanation: "In 1962 or 1963 – I don't remember exactly – during a (run) in the centrifuge he developed excessive spasm of the stomach. He then disappeared from our ranks." As for my pictures of the young blond pilot who turned out to be Ivan Anikeyev, Nelyubov's partner in disgrace, Leonov had given this description of his fate: "He was removed from the team because of his general physical condition. That was, I think, in 1963."

It is virtually impossible to believe that Leonov had so completely forgotten the scandalous expulsion of the arrogant Nelyubov and innocent Anikeyev; rather, he made up an innocuous cover story in the expectation that the facts would never come out to embarrass him. Such had been his orders.

What was happening?

With forty years hindsight, the 'big picture' has emerged, and here's what we now know:

1. To a degree unprecedented in any other space programme but fully in keeping with Soviet practice in pictures of all other subjects (particularly politically sensitive subjects), the USSR practiced deceptive photographic retouching of imagery in the early years of the Space Age.
2. To a similarly unprecedented degree Western publication and criticism of such forgeries forced official responses, some of them to the benefit of accurate public knowledge of events. Only the collapse of the Soviet Union brought real disclosure – a consummation we never really expected in the frustrating years when we wondered what had happened to all the erased men.
3. To the satisfaction of history, it was the Russians themselves (courageous journalists, crusading amateurs and debt-repaying purged officials) who ultimately provided the full and true history of these events – as it always should have been. A handful of Western enthusiasts had begun the process by focusing attention on the issue and by being the first to collect interviews and documentation.

As befits any field of historical study, analysis defined a sequence of discrete phases of this process:

Phase 1: 1957-1971 – There was limited imagery of space activities, the images were often mislabelled, and this led to widespread and often exaggerated Western scepticism of claims.

Phase 2: 1950s and 1960s – Common presentation in the Soviet press of Western space-related illustrations retouched to allege or imply Soviet origin.

Phase 3: 1971-1975 – Significant production and publication of retouched imagery, mainly of cosmonauts but also of hardware and history. There was a lot of Western speculation about the different reasons for different types of forgeries. Some reflected general propaganda themes or internal Kremlin power politics. Many were purely aesthetic, to unclutter or 'prettify' scenes. But many clearly continued to be motivated by historical deception for political propaganda purposes.

Phase 4: 1976-1986 – Under pressure from Western attention (and mockery), half-hearted and half-baked (and occasionally half-true) responses appeared in the Soviet media and in official interviews, but official lying still dominated.

Phase 5: 1982-1985 – "False dawn" and retrenchment. Following Brezhnev's death there was a marked relaxation of press restrictions on discussions of spaceflight problems by courageous journalists such as Yaroslav Golovanov. However, all such publications ceased with the accession of Chernenko.

Phase 6: 1986-1991 – The *glasnost* era under Gorbachev saw Soviet journalists publishing their own long-suppressed reports and photographs. 'Erased' cosmonauts were explicitly named in *Izvestia*. This was the only way, officials realised, to defuse the mocking laughter and to counter the well-earned disbelief in Moscow's official statements about their space activities.

Phase 7: 1991 to present – Almost full access to records and memoirs, public disclosure and non-restricted interviews. But recently there has been some retreat from this openness.

'X-2' IDENTITY REVEALED

It was in 1986, on the occasion of the 25th anniversary of Gagarin's flight, that we learned that cosmonaut 'X-2' really had screwed up royally. His name was Grigori Nelyubov. In Russian that actually means "unloved" and that sure was how he wound up. Officials feared that what he had done was so shameful that if the Russian public ever knew a cosmonaut had acted that way, the prestige of the successful heroes and the entire country would be tarnished. Nelyubov's sin wasn't financial, sexual or even particularly felonious. It came down to arrogance and selfishness – or at least an excessive level of those personality flaws. In Russian it is referred to as 'gusarstvo' or 'hussar-hood', the arrogant aggressive attitude of the medieval 'hussar' cavalrymen.

One early version of the story went like this. While returning to the cosmonaut training centre one night after a weekend in Moscow, Nelyubov and two other cosmonaut trainees were stopped by a security patrol. Their passes had expired but Nelyubov tried to bluff his way through with the claim, true as it turned out, that he and his friends were carousing cosmonauts. As such, Nelyubov insisted they didn't have to submit to any patrol of peasant draftees. Yet regulations were regulations and the soldiers would not back down. Nelyubov then knocked one down and began to push through the group towards the gate he intended to enter. The others grabbed at him and during a brief brawl the three cosmonauts were subdued. It didn't take long for the lieutenant in charge of the patrol to telephone the cosmonaut centre and verify the identity of their three captives. That was all it took to guarantee they would be released. Nelyubov was almost home free – and the universe awaited him. Spaceships needed flying and flying needed spacemen, and he was ready.

Expelled cosmonauts

The patrol's commander made one reasonable request. Some of the kids had been roughed up pretty badly in the brawl. He asked that Nelyubov and the two others – Valentin Filatyev and Ivan Anikeyev – sign autographs and apologise to the soldiers who, after all, had been doing their duty. Nelyubov's companions readily agreed but Nelyubov, possibly vindicated in his mind by the phone call that ordered his release, haughtily refused. He knew the officer had to let him go anyway.

What he didn't know was that the officer also had to file a report of the incident – although had he wished he could have forgotten the incident and put nothing down on paper. But in the light of Nelyubov's uncompromising attitude, the officer filed a full report. When this reached the director of the cosmonaut training programme – a stern 'Hero of the Soviet Union' who took very seriously their importance as role models for young Russians – Nelyubov was canned, and in a twist which fanned resentment towards him among the other cosmonauts his two companions in the brawl also were expelled. "They burned down together," one cosmonaut recalled years later, reviving the bitterness that he and his former colleagues still harboured for Nelyubov's poor judgement.

Later versions, based on more direct witnesses, differed in details but contained the same essentials. Ace Soviet-era space historian Yaroslav Golovanov, whose best works could not be published until the mid-1980s 'thaw' that preceded the USSR's total collapse, said the drinking bout occurred in a bar at the train station for Chkalov Air Force Base, about four kilometres from the cosmonaut village. He reported no fisticuffs but Nelyubov's refusal to apologise for the incident remained the proximate cause of higher-ups learning of it and dismissing all three men.

By the time his story came out, in the late 1980s, Nelyubov had been dead for two decades. Transferred to a jet squadron near Vladivostok, he sank into depression and alcoholism when his appeals for reinstatement were finally rejected, and he stepped in front of a train. More of that, shortly.

A famous photo session

A particularly curious feature of the retouched cosmonaut photographs of this era was that most of them originated from a single photo session in May 1961 at a resort in Sochi on the Black Sea. Following Gagarin's triumphal flight, the entire cosmonaut group was sent off for a well-deserved vacation. Whilst there, a series of sittings were conducted for a large number Soviet photographers, journalists and official historians – about a dozen, judging from the different viewing angles.

The order of the sequential views is unknown but one can speculate that the first shots involved all the cosmonauts present, with some training officials and a few family members – and Sergei Korolev, head of the space programme. Then there were shots of Korolev and the six cosmonauts assigned to Vostok mission training. Lastly were scenes of Korolev and Gagarin alone, both with and without Gagarin in full dress uniform. Numerous retouched Korolev-Gagarin images were later released, mostly altered to elide background details. No 'historical revisionism' or other fraud was evident.

The 'Sochi-6' photograph

The first-discovered Nelyubov erasure was from the 'Sochi-6' image and had three distinct versions with filled-in backgrounds for his missing body. There is also a second pair of before-and-after views from an oblique instead of a direct-on angle, with two different background fill-ins. And another independent deletion was found where a parachute instructor named Nikitin (who later died in a jump) was erased but all the cosmonauts were left in. The full-team photographs took years longer to come out, since there were so many unflown men. The last to leave the cosmonaut corps without flying was in 1969, for medical reasons. But even in this scene, one 'super deletion' forgery was found in which *five* unflown cosmonauts were carefully erased from among the crowd of ultimately flown cosmonauts.

Other faces erased

Two other scenes gave rise to additional cosmonaut erasures. Both dated to the first year of training, 1960-1961. I call the first scene the "parachute bus" as it involves a dozen cosmonaut candidates (including Gagarin preparing for a parachute jump) in front of a bus. The famous close-up of Gagarin in a flight hat is from this setting. I'm sure every trainee there had similar singleton portraits taken, but the group shot was the one that was doctored.

In the forged version there is something wrong with the pairing of trainees, as half of them stand in line while an equal number check out their parachutes. For example, Gagarin stands in front while Volynov bends over Gagarin's parachute from behind. Popovich is to the right of Gagarin and Nikolayev is to his right. But Popovich has nobody checking his parachute and the space between him and Nikolayev has been clumsily filled in with brush strokes. And Nikolayev's parachute checker is unknown because the photograph is cropped right down past his left ear – concealing even the existence of a checker. Years later I found a version of this picture that *did* include Popovich's parachute checker, but his face was turned downwards and I was unable to recognise him. Later, a Russian journalist identified him as Anikeyev – one of the two drinking buddies of Nelyubov who had been expelled.

As for Nikolayev's parachute checker, it took yet another discovery to identify him and the reason for his being cropped out. This time, printed in reverse but clearly the same shot, it was recognisably Bykovsky. An honourably flown cosmonaut (no shameful secret there) but he wasn't actually checking the parachute – he had stepped away and turned towards the bus tire. His upper body posture is suggestive of the reason for his removal. With his head bent low, his shoulders hunched forward and arms extended downwards, he looks like nothing so much as a man urinating on the ground. Whether that really was what took him out of line, even looking like he was peeing was sufficient to require his deletion.

Iconic image airbrushed

The second multi-deletion scene is on the bus to the pad for Gagarin's launch on 12 April 1961. A still from a 16-mm movie shows him sitting in his spacesuit, with his similarly suited backup Titov sitting directly behind him. Two other cosmonauts, in uniform, stand behind Titov's seat – Nikolayev and Nelyubov. Viewed from in front, Nelyubov's head is recognisable behind Gagarin's helmet. The first version, released as early as 1961 (even before Nelyubov's expulsion) has his faced smeared out to conceal his status as an unflown cosmonaut. Other versions, in indeterminable order of their creation, have his head removed and a fake background painted in – but his uniform jacket and collar remain. The next step is a view in which even that trace of him is removed with clumsy fill-in forms, until finally the entire background behind Gagarin is blacked out.

Significantly, this image of a smiling Gagarin on his way to his launch, with the faked backgrounds, is probably the most popular of all images of the world's first cosmonaut on his launch day. Officially it was emblematic of the USSR's glory, but in hindsight it also inadvertently symbolised the factual photographic falsifications that characterised official information in the Soviet era.

THE TRUTH ABOUT NELYUBOV

Although an outline of the fates of the missing cosmonauts was first revealed by Yaroslav Golovanov in *Izvestia* in 1986, details remained sketchy. The sea-change in Russian space history studies culminated in a 2007 documentary film about the fate of Grigori Nelyubov. Entitled *Он мог быть первым* (*He Could Have Been First*), the 45-minute video featured lengthy interviews with principal players in the drama, including his widow Zinaida and his younger brother Vladimir.

The first third of the programme goes into personal details of his youth, his flight training, his romance with his new wife, and the cosmonaut programme in which he was one of three candidates for whom special form-fitting spacesuits were assembled

Nelyubov on bus to Vostok launch, several different forged variations [Apr 12, 1961]

Upper left: original still
Bottom: Three variations
Right: Most extreme variation

4.3: An iconic image of Gagarin was 'edited'.

for the first flights. By the middle of the programme it is late 1961. Nelyubov is next in line to fly, but then political factors intervene and other cosmonauts are assigned to special symbolic missions. While still in training for a near-term mission, Nelyubov is injured in a centrifuge and placed on medical leave with reduced duties.

In the spring of 1963, Nelyubov had a nervous breakdown and was taken entirely off duty for three weeks. Free of active duty restrictions, he decided to go out for a beer because alcohol was not allowed in the cosmonaut village. At a nearby military airfield off-base bar he met two other cosmonaut trainees – Anikeyev and Filatyev. Being off duty, he was in civilian clothes. According to Zinaida, the two uniformed cosmonauts began a noisy 'hand speed' competition at their table which became so rowdy that a military patrol was called. All three were taken in for questioning and it was then that Nelyubov, still recovering from his nervous breakdown, exhibited his 'hussar-ness'.

Attempts to save cosmonaut career

The next day, the political officer of the cosmonaut team visited Nelyubov at his apartment and said a formal apology was in order to prevent the report from going forward. Zinaida heard him promise to do so, but he never made the apology.

General Kamanin, head of the cosmonaut programme, got the report the next day. In his diary entry for that night, he wrote that the other two cosmonauts were "of no value", with previous behaviour problems and low academic performance. General Kamanin, with Gagarin's concurrence, expelled them immediately. But Nelyubov, Kamanin wrote, was another matter. He was one of the top cosmonaut trainees. He'd been in civilian clothes, off duty and, by the patrol's report, had attempted to get his rowdy mates to leave. The consensus was that some way must be found to save his career.

At a political meeting organised by cosmonaut Pavel Popovich, Nelyubov was to listen to a round of condemnation of his behaviour and then apologise to his fellow cosmonauts. But it didn't work out that way and he was expelled. His wife recalled that she cried, and that Nelyubov was calm. "Well, it's not the end of the world," she remembers him telling her.

But the end of the world was almost where they were sent – the Pacific coast of Siberia, at an airfield in the middle of the forest with only a few buildings for the military personnel. Water was drawn from wells. They were 50 kilometres from the nearest town. Their furniture from Moscow took six months to catch up with them. Yet, to his wife's surprise, Nelyubov bounced back. His flying skills and cheerful nature returned. He seemed to be settling into and thriving in his new life. He made new friends and became "the darling of the regiment" but then his cosmonaut friend Pavel Popovich, making a tour of the region, visited the city of Khabarovsk and told local officials that he had a friend at the air base.

Nelyubov was flown in right away and for a day and a half he was a 'cosmonaut' again. Even though Popovich had a one-on-one conversation to attempt to reconcile Nelyubov to his fate, their reunion had the opposite effect. He wanted back in. "The meeting only harmed Grisha," Zinaida recalled many years later. "Somehow he was stirred up again – his mood became terrible."

Another rejection

Nelyubov flew to Moscow and visited the cosmonaut centre. Kamanin had apparently promised that he would only have to spend a year or two in exile, and would then be allowed back. Even Gagarin promised to push for his reinstatement and so Nelyubov returned to Siberia in high hopes.

The next two years were full of ups and downs, as Nelyubov applied for advanced training and transitioned to the new MiG-21 interceptor. While in training at Lipitsk he met with Marina Popovich, the cosmonaut's wife and a test pilot in her own right. She, too, was encouraging, and worked to get him into test pilot school. The school asked for his personnel file and a transfer seemed imminent. Friends recalled him at this time living "in suitcase mode" (packed up and ready to go) but then he received a curt telegram turning him down due to some "reorganisation". He was thunderstruck.

After a flurry of letters, Marina Popovich told him the true reason. A cosmonaut – even in the 2007 documentary interview she still wouldn't say which one it was – had warned the selection board about Nelyubov's 'hussar-ness', saying: "If you want to get into trouble with his audacious (дерзкий, also implying 'impudent' or 'cheeky') character, then take him into the tester programme."

Doomed finale

Devastated by the rejection and its secret cause, Nelyubov began drinking in earnest. He would visit the local train station, mix with passengers in the cantina and regale them with cosmonaut stories and autographed photographs of his former colleagues while they bought him round after round. Nelyubov roused himself for one further effort: an appeal directly to Sergei Korolev. Korolev had always favoured Nelyubov owing to his sharp academic skills, and according to some historians Nelyubov had been Korolev's choice for the first mission. "He knew that Korolev was well inclined towards him," said Zinaida.

In making plans for another trip to Moscow, Nelyubov was again "on cloud nine" and out-going. He was the master of ceremonies at the Officers Club party for New Years 1966 and didn't even need to drink champagne in order to feel joyous. In what everyone took to be a good omen, he won the door prize – a plastic cosmonaut figure. Fellow pilot Vladimir Upyr recalled, "Grisha cheered up. He was another person. Twinkle in his eye."

Two weeks later the newspapers announced the death of the 59-year-old Sergei Korolev during surgery. Nelyubov's last hope was abruptly crushed and he smashed the plastic cosmonaut figurine to pieces and sank into dark depression. A few weeks later, his wife repeated what had become a routine protective measure in the evening. She locked the front door of their top-floor apartment and went to spend the night with friends, leaving Nelyubov inside without alcohol or other threats. However, on the night of 17 February he got out, probably by climbing down the outside balcony, and walked to the train station, onto the tracks and right into a passing locomotive.

Zinaida's taped comment is heart-breaking in its pain. "For some time I had been afraid that this could happen," she said, biting her lip and looking off camera. She knew it had been deliberate because he had left her a note, which she read for the camera. "You were always the best of all. One needs to really search for such a woman. Forgive me." He hadn't signed it.

Did he know he was 'erased'?

There is one puzzling aspect of Zinaida's interview shown in the *He Could Have Been First* documentary. Apparently, according to the narrator, Nelyubov kept a clipping of a cosmonaut group photograph that had been published in *Izvestia* just after Gagarin's flight. "The picture shows the whole group," the narrator intones. "In it, there is no cosmonaut Nelyubov." Zinaida's words were explicit: "He said: 'I already don't exist.' He has already been deleted from the lists, and from all of the photos they even have retouched his image."

This is a poignant story but there is one major problem with it: I cannot believe that any mid-1960s-era Soviet newspaper would have published a photograph with as-yet-unflown cosmonauts. This would have been contrary to the strict censorship practice

4.4: Nelyubov is cropped from history. A 1961 view of Gagarin and Titov in Red Square, and a 1980s release showing them talking to an erased man.

that had nothing at all to do with Nelyubov's personal failings. Yet Zinaida Nelyubova told the interviewer that at least one of the scenes had been published in *Izvestia*.

At least two scenes with Nelyubov 'gone' were published prior to his death. The first is him behind Gagarin on the bus to the launch pad. I have seen that in a 1961 Soviet military newspaper from shortly after the flight, with Nelyubov's face crudely smudged out. The second is the scene of Gagarin and Titov talking in Red Square but with Nelyubov cropped out of the right side of the photograph, published shortly after Titov's flight. Both these cases were consistent with the general Soviet of policy of not showing faces of any unflown cosmonaut. And since they were published while Nelyubov was still a cosmonaut in good standing (he was not expelled until 1963), he would have seen them then, and hence could not have been alarmed by the deletions of his own likeness.

I suspect Zinaida's comments are retrospective mnemonic editing of the narrative of her husband's horrible fate. Otherwise, I can find no justification to believe her statement that her husband had seen his own face being punitively deleted from the cosmonaut programme.

Zinaida had also managed to play the game herself by getting Grigori's picture in front of the Russian public. After returning to Moscow she made the acquaintance of Svetlana Savitskaya, daughter of the Soviet defence minister and an avid aerobatics pilot and skydiver. They apparently became friends and years later, after becoming a renowned cosmonaut herself, Savitskaya wrote her autobiography. In describing her pre-cosmonaut flying activities, she included photographs of many colleagues, and one of them was of a smiling Zinaida, identified as "sports parachutist" along with a second happy character identified as "military pilot Grigori Nelyubov" – but with no mention of his cosmonaut career.

COSMONAUT DEATH COVER-UP

Not all of the cover-ups of the 'unflown' cosmonauts involved photographic forgeries – one actually involved a falsified grave stone. Another man ('X-7' on my original 1973 roster of 'unknown unflowns') took a secret to his grave that might have given NASA a fighting chance to save the lives of the Apollo crew who perished in the pad fire in January 1967.

Valentin Bondarenko was killed in a pressure chamber fire weeks before Gagarin made the first-in-history human space flight. His death led to the notorious rumours that Moscow had lost "unidentified cosmonauts" on "secret space missions". The fire was extremely ferocious because it involved enriched levels of oxygen and a careless abundance of flammable materials – a warning that NASA might (or perhaps might not) have taken to heart before having to learn it through their own mistake.

"Valentin was the youngest of the first batch of cosmonauts," wrote Yaroslav Golovanov in his bombshell 1986 *Izvestia* article. A small, grainy formal portrait accompanied the article. It showed a young man of 24 attempting to look stern and important. The photograph had been taken only a few days before his death.

Bondarenko had been undergoing routine training in a pressure chamber, which was part of a ten-day isolation exercise. At the very end of the exercise he made a trivial but fatal mistake. "After medical tests," explained Golovanov. "Bondarenko removed the sensors attached to him, cleaned the spots where they had been attached with cotton wool soaked in alcohol and without looking threw away the cotton wool – which landed on the ring of an electric hot plate. In the oxygen-charged atmosphere the flames immediately filled the small space of the chamber."

Under such a condition of high oxygen concentration, normally non-flammable substances can burn vigorously. His training suit caught fire. Unaccustomed to the vigour of high-oxygen fires, Bondarenko would only have spread the flames further by attempting to smother them. When the doctor on duty noticed the conflagration through a porthole, he rushed to the hatch, which he was unable to open because the

Valentin Bondarenko [1937-1961], first cosmonaut training fatality. From left: "X7" unknown cosmonaut from 1960 training scene [Oberg, 1973]; Bondarenko in 1960 parachute training group shot; formal military portrait; Valentin with wife Hanna ['Anya']. He was killed before the Sochi photos.

4.5 Cosmonaut Valentin Bondarenko.

internal pressure kept it sealed. Releasing the pressure through bleed valves took at least several minutes, throughout which Bondarenko was engulfed in flames. "When Valentin was dragged out of the pressure chamber," continued Golovanov, "He was still conscious and kept repeating, 'It was my fault, no one else is to blame.'"

Bondarenko died eight hours later from the shock of the burns and was buried in Kharkov, in the Ukraine, where he had grown up and where his parents still lived. He left a young widow, Hanna, and a five-year-old son, Aleksandr ('Sasha'). Hanna, or 'Anya' in the Russian version, remained at the cosmonaut centre in an undisclosed job. When he grew up, young Aleksandr became an air force officer and also worked for many years in cosmonaut training.

Golovanov's candid disclosure of Bondarenko's death may have astonished his countrymen (and it briefly made headlines in the Western press) but it was hardly news to informed space sleuths because they had been hot on the trail of this very incident and Soviet censors knew it. The cause and effect of Western digging into a Soviet catastrophe, followed by Soviet large-scale (but still not full-scale) release of an "official account" are clear-cut. The broad outlines of the "Bondarenko tragedy" had already slipped past the Soviet cover-up.

Russian rumours reach the West

In 1982 a recently emigrated Russian Jew named S. Tiktin discussed Soviet space secrets in a Russian-language monthly magazine published by anti-Soviet émigrés in West Germany. He mentioned almost in passing a relevant incident. "Soon after the flight of Gagarin [in 1961] the rumour spread about the loss of cosmonaut Boyko (or Boychenko) from a fire in a pressure chamber," he wrote. This is clearly a garbled version of "Bondarenko". In 1984, St. Martin's Press published the book *Russian Doctor* by the Russian émigré surgeon Dr Vladimir Golyakhovsky. He described the death of a cosmonaut trainee in a pressure chamber fire. He had apparently been the emergency room doctor at the prestigious Botkin Hospital when the man was brought in, and he devoted half of an entire chapter to the incident.

As Golyakhovsky recalled it, a severely burned man identified only as "Sergeyev, a 24-year-old Air Force Lieutenant," was brought in on a stretcher. "I couldn't help shuddering," Golyakhovsky recalled. "The whole of him was burnt. The body was totally denuded of skin [and] the head of hair; there were no eyes in the face ... It was a total burn of the severest degree. But the patient was alive ..."

Golyakhovsky saw the man's mouth moving and bent down to listen. 'Too much pain – do something, please – to kill the pain' were the tortured words he could make out. "Sergeyev" was scorched everywhere except for the soles of his feet, where his flight boots had offered some protection from the conflagration. With great difficulty the doctors inserted intravenous lines into his feet – they couldn't find blood vessels anywhere else – to administer painkillers and medication. "Unfortunately, Sergeyev was doomed," the surgeon remembered realising immediately. "And yet, all of us were eager to do something, anything, to alleviate his terrible suffering."

Afterward, Golyakhovsky spoke with a small young officer who had waited by the phone in the lobby. The doctor sought and received an account of the accident, which involved "an altitude chamber ... heavily laden with oxygen" and "a small electric stove (with) ... a rag burst(ing) into flame". He was also told that it had taken half an hour to get the pressure chamber open, with "Sergeyev" on fire until the flames had consumed all the oxygen in the chamber. Sometime later, Golyakhovsky wrote, he saw a picture of this officer in the newspapers and recognised him as Yuri Gagarin. But we now know what Gagarin was doing during those final days before his Vostok launch, and it was not attending a dying comrade. It was probably another short blond cosmonaut, and perhaps Anikeyev.

Despite minor distortions, the Tiktin and Golyakhovsky material turned out to provide fundamental, direct, and invaluable leads into a major catastrophe in the early Russian space programme. It was left to the Soviets only to fill in the details about the real death of Valentin Bondarenko – which they did in April 1986.

Missed chance for Apollo astronauts?

When Bondarenko's tragic story was taken up by the Western press, I was widely interviewed for historical background. In response to this revelation about his fiery death, and about the young wife and infant son that he had left behind, I received a touching letter from an American who fully sympathised with their loss. It was from Don Chaffee, whose 31-year-old son Roger had been killed in the flash fire which hit Apollo 1 in January 1967. Chaffee asked me how he could send Bondarenko's family a copy of the book that he had written about his son's life. The obvious route was to contact Golovanov. In July 1988 Golovanov published an article in *Izvestia* about what happened next. He forwarded the letter from Chaffee, with my translation of it, to Hanna Bondarenko's home in Kharkov and then telephoned her to seek permission to release her address.

"I am very grateful to Mister Chaffee for his warm letter," she told Golovanov as she authorised the exchange of addresses so that she could be sent a copy of the book and then she added, "Roger Chaffee and my husband Valentin Bondarenko arrived in cosmonautics in the period of its youth, and they themselves were young men full of strength and glorious plans. Well, any great affair involves losses, sometimes human losses. Roger and Valentin did not fly into space but they gave their lives happy that in so doing their comrades could fly there."

Golovanov expanded on this theme in the newspaper article. "Reading the letter of Mr. Chaffee and hearing Hanna," he wrote. "I thought about how it's not just smiles that bring people together but grief is able to accomplish it too, if people believe in thoughtfulness and good will." From these exchanges, over the following ten years the two families developed a friendship that ended only with Don Chaffee's death in 1998.

One final revision was required to convert 'Soviet reality' into a more truthful version of reality. The original headstone on Bondarenko's grave only contained a brief tribute from his "fellow pilots". Sometime around 1990 an additional line was engraved, adding "cosmonauts of the USSR". Sometimes, retouching serves truth after all.

4.6: These photos of Bondarenko's gravestone by Bert Vis show the recent addition of the words 'Cosmonauts CCCP'.

OTHER DELETIONS, OTHER MOTIVES

Even when the disqualifying reasons were not dishonourable or disastrous, the Soviet passion for the appearance of perfection compelled their censors to perform similar photo forgeries. Ten years after Gagarin's flight, one rookie cosmonaut was dropped from a launch for purely medical reasons – but relating to another cosmonaut, not himself. So when photographs of the other cosmonauts were released later on they were altered to mask his presence in the group.

The most striking before-and-after forgery involving his face is called the 'On Top of the World' shot from early 1971. It is part of a sequence showing the first three Salyut 1 space station crews gathered around a giant globe. This scene was released in 1972 in the documentary *Steep Road to Space*, dedicated to the tragic climax of the Soyuz 11 mission in June 1971. The figures were recognisable as the two crews that had flown to the station (Shatalov, Yeliseyev and Rukavishnikov on Soyuz 10 which was unable to dock properly; and Dobrovolsky, Volkov and Patsayev, who reached the station successfully but died during their return to Earth), in addition to veteran cosmonauts Leonov and Kubasov and a ninth unknown figure.

A year later, when Leonov and Kubasov were switched from the stalled Salyut programme in order to fly the Apollo-Soyuz linkup of July 1975, photographs were released of them in a Soyuz trainer. But several of the pictures showed indications of

ON TOP OF THE WORLD
A vanished Salyut 1
cosmonaut crewman

Published still of celebration
as 'Rukavishnikov joins the
[first] crew' -- but a later
version [right] shows another
rookie cosmonaut LEAVING
the group of candidates.
NOTE ALSO: Under Volkov's
chin [far left], medals on
Leonov's uniform, removed

4.7: Pyotr Kolodin vanishes.

a third crewman training with them – specifically portions of his body at the edge of the frame. The most striking scene shows the mystery crewman's wrinkled forehead and part of his cap in the extreme lower right corner of the shot.

This, as we learned years many later, was Pyotr Kolodin – and that name, like Nelyubov's and Bondarenko's and others – never made it onto any roster of 'flown cosmonauts'. Not even selected until 1963, he had not been present for the Sochi photographs and therefore had not been a target for the associated deletions. Leonov, Kubasov and Kolodin had been within days of blasting off to become the first men to occupy Salyut 1, but a medical problem with one of the cosmonauts had grounded the entire crew. Their backups went up instead and when an air leak struck the returning ship three weeks later they died instead of the men who had, until that moment, been cursing their bad luck at having been grounded by what proved to be a medical false alarm. The Soyuz was modified, the crew reduced to two cosmonauts who now wore pressure suits, and the third seat replaced by emergency oxygen supplies. Leonov and Kubasov were assigned to Apollo-Soyuz, and Kolodin, without a flight assignment, failed his next flight medical exam and was grounded.

Several years later the April 1971 pre-launch group photograph was released with Kolodin airbrushed out. Because he was at the extreme right he could readily have been cropped out; the need to airbrush him is unexplained. Curiously, and apparently for aesthetic reasons, the visible parts of another uniformed cosmonaut on the left of the picture were also airbrushed out. Wider views of the same scene in

4.8: The original Soyuz 11 crew images included cosmonaut Kolodin (right).

Steep Road to Space identify him as Leonov. Hardly a 'disgraced' cosmonaut, but in this case a visual distraction.

Kolodin, no longer a cosmonaut, remained in the space programme as a ground operator. And twenty years after he had come within days of launching into space, and a decade after I had published a set of before-and-after pictures that showed him being erased from official histories, I was introduced to him on the floor of Mission Control in Moscow. He was amused by his notoriety (a moral victory over officials who had tried to "unperson" him), and we shook hands warmly. To actually touch an 'erased cosmonaut' was an immensely satisfying experience that I'd never expected, or even dared hope for. It was more than adequate compensation for the extinction of what for years had been an enjoyable hunt for more examples of such forgeries. In post-Soviet Russia, the practice of erasing the images of failed cosmonauts has itself, at long last, been erased.

More forgeries discovered

It would be too simplistic if all cosmonaut photograph forgeries could be traced to the single motivation of concealing the existence of unflown candidates. The search for before-and-after pairs turned up other examples of forgeries where the motivations were much less clear cut.

Aesthetics
(& peace propaganda)

Korolyov bids farewell to Gagarin at launch pad, and extra characters are one-by-one removed so false fill-in is needed (but fence stops). Marshall Kiril Moskalenko, missile commander, departs.

4.9: Aesthetics versus propaganda.

4.10: A man's face disappears from another Soviet-era image.

In one multi-step forgery sequence from 12 April 1961, Sergei Korolev is seeing off a spacesuited Yuri Gagarin on the launch pad at Baykonur, with Soviet military rocket commander Kirill Moskalenko standing with them. In the original scene there are other pad workers in the background, but a 'cleaned up' version elides the small figures and leaves the three main players alone. Then the major forgery occurs, with Moskalenko (Chief of the Strategic Rocket Forces and successor to Nedelin, who was burned to ashes in a missile disaster less than six months earlier) being erased and a false background painted in. Another version that masks Moskalenko's presence has Korolev's left arm raised to grasp his hat, totally covering the military officer. I can't say whether this was a convenient accidental shot or an artistic concealment tactic.

On the bus to the pad for Voskhod 2 in March 1965, spacesuited cosmonauts Belyayev and Leonov sit in front and cosmonaut Komarov kneels between them, deep in conversation with mission commander Belyayev. Sitting behind Leonov and peering intently over his shoulder there is another uniformed officer – at least in the original image. The photograph actually published soon after the 1965 flight showed only an empty space where that man's head had been. Curiously, another version (the only copy of which I have is a poor-quality transparency from the private collection of a journalist) shows that same head partially concealed behind Leonov's seat.

What did it all mean?

In assessing what these photograph forgeries meant, it can be helpful to enumerate what they did *not* mean. And there has been no shortage of misinterpretations and misrepresentations.

First, the forgeries are *not* evidence for secret Soviet space fatalities – in fact, just the opposite. In the early 1960s there were many Western press stories alleging that cosmonauts had been killed in secret space missions. Some of the stories were even marginally plausible – at least at first – even if the evidence for them was weak or ambiguous. But by the time the forgeries became widely known, the stories were already discredited by the same sort of archive revelations that produced many of the forgery before-and-after scenes. And the full records of the selection and training of the first cosmonaut team showed there were no 'missing cosmonauts' unaccounted for. Some did disappear from photographs, but their fates were well documented and none involved spaceflight accidents.

Second, the photographs were *not* evidence for the most persistent of the Gagarin-era rumours that senior Soviet test pilot Vladimir Ilyushin actually was the first man in orbit – a story which is periodically resurrected by cable channel documentaries. Amusingly, the Sochi-6 before-and-after cosmonaut group forgery is featured on an Internet website promoting the Ilyushin-was-first theory and other dubious claims made by the Italian radio-listener brothers Achille and Giuseppi Judica-Cordeglia. Less amusingly, the website claims 'copyright' of the Sochi-6 photographs.

The real question of copyright was a genuine issue when I published *Red Star in Orbit* in 1981. Lawyers from the NYC publisher, Random House, wanted assurances that we had the legal right to use such images. Fortunately, I was able to establish to their satisfaction that since they were official Soviet government products, and since that government had never signed the International Copyright Union convention (and in fact enthusiastically violated thousands of foreign copyrights of books, music, and other intellectual properties), the images were legally owned by nobody. They were – and despite any modern overreaching websites, still are – free of copyright restraints.

Third, the deletions are *not* related to political differences such as the notorious Stalin-era revisions. One or two later cosmonaut candidates did get into trouble over inadequate political correctness, but there is no evidence they were erased from any photographs; they were simply fired.

4.11: Arthur C. Clarke cheekily gave friend Alexei Leonov a copy of Oberg's book during a 1982 visit to Moscow.

Fourth, contrary to explicit excuses that Soviet diplomats in Washington made to me in the mid-1970s, they were *not* interlopers who sneaked into the scenes without permission and were then chased off before taking the group pictures again. I'm not making this up – that was the straight-faced story that I was told.

Lastly, they were *not* kidnapped by space aliens – another version that has long festered in darker corners of the Internet. Based loosely on fables first circulated by 'journalist' Frank Edwards in the 1960s, these stories of alleged disappearances in space don't even try to explain why, even if true, such events would then require the same victims to disappear from archival images.

The continuing anonymity of unflown cosmonauts from later classes continued as official Soviet policy right up until the final days of the USSR. It wasn't until 1989 that the full 'unflown cosmonaut' roster became public, purely by accident. I was at Star City consulting for a *60 Minutes* news team doing a programme on cosmonaut training. While the crew was being videotaped doing exercises in the gymnasium, an incredibly careless escort officer directed me to get out of the scene and "wait over there" – and took me through a door into the locker room. Left alone there, I quickly noticed that all the dozens of lockers had individual names posted on them. As calmly but as quickly as I could manage, I discreetly jotted down every name up and down four rows. But that's another story!

The last 'unknown'

Perhaps the greatest 'unknown' about these forgeries, in retrospect, is not why they were performed but *who* ordered them? As far as I can tell, there is no 'paper trail' leading back to any specific person. At least, I haven't found it yet.

But one approach might be to develop the characteristics we can presume the perpetrator would have had – even if we don't know who that perpetrator was. In keeping with the designation scheme of the X-cosmonauts from forty years ago, for convenience I want to label this man as 'X-0', the cosmonautical cipher.

To perpetrate the fraud, X-0 must have been highly placed within the cosmonaut programme during the period that those forgeries were made, specifically 1960-1971. And in order to know who needed to be erased, he must have known by face all of the cosmonaut candidates. Furthermore, he was probably familiar with the Stalin-era mass photographic falsifications carried out for national propaganda purposes, so he couldn't have been too young. He would also have been somebody who was very concerned with public image of 'Soviet heroes' – and he might even have been an official 'Hero of the Soviet Union' himself. X-0 might have been at the very top of the cosmonaut programme. He may have been personally involved in making the real cosmonaut candidates "disappear", and eliminating them from the record was just a matter of tidying up.

There are additional clues to be found in the differences between the 'cosmonaut forgeries' and typical Soviet falsifications. Perhaps these can also lead to identifying X-0. The biggest difference is that unlike the run-of-the-mill political forgeries of the 1930s and later, which removed people who were already well-known to the public but had become 'bad guys', these forgeries removed people who were unknown to the general public – in order to keep them unknown. This is a very important clue, as the number of people who did know who these men were, which included X-0, was rather limited. So candidates for the forger-in-chief are few in number. Some of the removed figures (such as Anikeyev in the parachute preparation line-up) were hard to recognise. Only somebody who knew all of the original figures by face, and probably personally, would know who needed removal. The decisions had to have been made at Star City, but detailed instructions could have been provided to existing forgery workshops run by the Soviet news bureaucracy. The artists who did the actual work probably didn't even know who they were erasing – and knew not to ask.

Identity of forger-in-chief?

When the list of characteristics is compiled this way, I think only one name rises to the top of the short list of possibilities for X-0. This is General Nikolai Kamanin (1909-1982), head of the cosmonaut programme from 1960 to 1971. Kamanin was the first officially designated 'Hero of the Soviet Union' and was later a military hero of the Great Patriotic War. Straight-laced, strict and merciless, he would become the cosmonauts' godfather, taskmaster, confessor, ultimate judge, jury and 'executioner'. The more one learns of him, the easier it is to imagine him protecting the 'hero myth' by marking up photographs with a red crayon for the forgers to go to work on.

Yet he was also honest with himself, in private, and kept candid diaries. They are full of detailed insights into the highs and lows, rewards and challenges, promotions and punishments, that he was involved with concerning the cosmonauts. For all that, they do not address the question of photographic falsifications for the sake of public patriotic imagery. But they do reveal a character who could easily find such methods justifiable – even necessary. Kamanin is such an obvious leading candidate for 'X-0', researchers must consider the possibility that the real 'X-0' concealed his own existence as thoroughly as he did those of the erased cosmonauts. One potential candidate is a mysterious figure, "KGB officer M. S. Titov", who appears with cosmonauts in three photographs from 1961. No further information is available, although a thorough Internet search located a WW-II military officer named "Mikhail Stefanovich Titov" [1923–1986] whose much-younger photograph bears a slight resemblance to the 1961 Titov. However, his official obituary made no mention of any KGB service or any association with cosmonauts (nor in Soviet times should it have been expected to). 'X-0' covered his tracks very well.

In perhaps the greatest erasure in the history of Soviet manned space flight, X-0, the man who ordered the erasure of so many of the early cosmonauts and their fates, met with only temporary success in these endeavours aimed at his victims. But regarding himself, as the author of the forgeries, X-0 took measures which so far have effectively erased his falsification role from history. Until that final erasure is remedied and the hole that X-0 tore in the fabric of cosmonautical reality is refilled with the whole truth, the mystery of the 'erased cosmonauts' will be a story without an ending.

5

The view from Paris

by Claude Wachtel and Christian Lardier

In France the study of the Soviet space programme was centred on the society Cosmos Club de France (C2F). Starting in the mid-1960s, it pioneered the art of studying the Russian-language books available in Western bookshops for obscure clues inadvertently passed by the Soviet censor.

The C2F was founded by French space journalist Albert Ducrocq in October 1963 to encourage young students, and many of its members would go on to participate in the French space programme. His interest in the USSR began with a visit to Moscow in the early 1950s when he was shown an early computer at the Institute of Precision Mechanics and Technical Computing, headed by Academician S. A. Lebedev. After the start of the Space Age in 1957 Ducrocq became a well-respected commentator in the French media, and invited members of the public into the studio to participate in his weekly radio show about events in space. It was during one such broadcast that a teenage Christian Lardier first encountered the writer.

One of the earliest events organised by the C2F was a showing of the Soviet movie *The Twins of Space*, about the double flight of cosmonauts Nikolayev and Popovich, at the Monte-Carlo cinema on the Champs-Elysées. Albert Ducrocq was also on hand to welcome Yuri Gagarin when he arrived to attend the week-long XIV International Federation of Astronautics (IAF) congress in Paris in 1963. At the end of that year the C2F formed a 'Soviet Section' under the direction of Alain Dupas and Alain Didier in order to study the history of Russian cosmonautics, and, starting in 1964, this group published its findings in the society's French-language journal *Orbite*. The magazine had a print run of about five hundreds copies and many of its scoops were picked up outside the country, but it was rarely quoted as a source by other sleuths.

5.1: The C2F's research team with founder Albert Ducrocq.

Finding secrets in Soviet books

by Claude Wachtel

My passion for space goes back to the spring of 1965 when I attended a meeting of the Cosmos Club de France in the broad lecture hall of the Ecole Supérieure des Techniques Avancées in Paris. Two years later I met Yuri Gagarin at the Le Bourget Air Show, and I will never forget the smile he gave me when I was queuing to get his autograph. Perhaps that is the root of my first professional engagement with the space field and my application for the first French-Soviet spaceflight fifteen years later.

FIRST STUDIES

My work on the identity of space "glavnye konstruktory" began in 1965 after seeing an article attributed to a "Professor G. V. Petrovich" (later identified as Academician V. P. Glushko) in the November issue of the magazine *Aviation and Cosmonautics*. This identified the functions of the major designers of rockets and satellites, cosmic engines, guidance systems, and launch infrastructure, as well as a scientist referred to as "the theoretician of the cosmonautics" (later identified as Mstislav Keldysh).

The secret nature of Soviet cosmonautics piqued my curiosity, and I began to buy issues of *Pravda* and translate its articles a word at a time using a dictionary. In 1969 I published my first article about Cosmos satellites in the journal of the C2F. Later in that year I travelled to England to see the receiving station of the Kettering Grammar School which listened to radio signals from Soviet satellites, and met Geoffrey Perry. The same year I started my scientific studies at the Pierre & Marie Curie University in Paris, and in 1975 obtained my doctorate in space geophysics.

Parisian bookshops

At that time only two bookshops in Paris specialised in Russian publications: the Globe and Dom Knigi (House of Books). The Globe was a modern establishment, with the part dedicated to Russian books controlled by "Madame Olga" – a fervent supporter of USSR who would later confide to me her indignation at seeing Stalin's decisions contested during the *perestroika* era. Starting in 1969, I ordered most of my Russian books from her. Dom Knigi was the exact opposite, being controlled by two very old men who spoke clumsy French and who were White Russian refugees with no interest in Russian culture. Their shelves reached the ceiling and were overloaded with a jumble of books. Customers using stepladders could sometimes find hidden treasures, including books from the 1950s. One day, Christian Lardier came back with two copies of a 1966 book by Vasily Mishin (Korolev's successor) at a time when his name was still very secret. Today I still preciously guard my copy.

Р. Ф. Аппазов, С. С. Лавров,
В. П. Мишин

Баллистика
управляемых ракет
дальнего действия

Издательство «Наука»
Главная редакция
физико-математической литературы
Москва 1966

5.2: Vasily Mishin's book from 1966 discovered in Paris.

During twenty years I accumulated almost one thousand books in my Avenue de Clichy apartment, but have since donated most of them to Christian. I retained some, mainly those relating to major space events or written by chief constructors like Viktor Makeev, Semyon Alekseev, Vladimir Chelomey, Grigori Voronin and others. At that time, I realised that an amazing amount of information about so-called 'secrets' could be obtained from these documents.

It was the diversity and quantity of these Russian-language documents collected by us that enabled the French space sleuths to publish the first complete story about many major space events – not only in comparison to other Western researchers but years before the information was finally published by the USSR.

Reading between the lines

The publication of the *Cosmos Encyclopédie* series edited by Albert Ducrocq gave me the opportunity to use these documents. At the C2F, I had become the "specialist" on the Cosmos satellites – the designation used to hide, as far as possible, the military role of many Soviet satellites. At that time, my postman delivered seven newspapers to me each day. After each Soviet launch, *Pravda* included a news release that was almost always the same. That word "almost" is important, because I learned to detect and interpret all those minor differences. I cut out each news release and pasted it on a card.

Уткин Владимир Федорович (р. 17.X 1923) — механик, акад. АН УССР (1976). Чл. КПСС с 1945. Р. в Пустоборе (ныне Рязанской обл.). Участник Великой Отечеств. войны. Окончил Ленингр. мех. ин-т (1952). Осн. науч. работы в области прикладной механики и машиностроения. Дважды Герой Соц. Труда (1969, 1976). На XXV съезде КПСС избран чл. ЦК КПСС. Деп. Верх. Совета СССР 8—10-го созывов. Ленинская пр. (1964).

5.3: The C2F identified Vladimir Utkin as Yangel's successor.

I also created a second file with the names of the article authors and the people quoted in the reports. I soon realised that those who were the most quoted were not necessarily the most interesting; often it was those who rarely appeared. In 1971 the signatories of the obituaries of Mikhail Yangel, Alexei Isaev and Georgi Babakin made it possible to obtain key information on the organisation of the space industry. These also identified the role of the 'Ministry of General Machine Building' as the Soviet equivalent of NASA; that is to say, the lead agency of the space industry, as well as eight other ministries of the military industrial complex that, in one way or another, were involved in the space effort. After studying a lot of these documents, I concluded that the chief constructor's name was V. F. Utkin – chief designer of the design bureau 'Yuzhnoe' in Dnepropetrovsk and others.

New sources of information

In 1971 I obtained a second source of information – the *Great Soviet Encyclopedia*, a yearbook that was first published in 1957. This contained short biographies of new members and correspondents of the Academy of Sciences. The names of Valentin Glushko, Viktor Kuznetsov, Sergei Korolev and Mstislav Keldysh were all present, but with brief descriptions that listed them as "specialists in mechanics". Thankfully the text mentioned their distinctions as Hero of Socialist Labour in 1956 or 1961, or Lenin Prize winner in 1957. At that time we already knew that Korolev (who died in 1966) was the designer of the famous Sputnik and Vostok, but Glushko was known only for his pre-war work on early rocket engines. If the real positions of the space specialists were hidden, the dates of their common awards (for example the launch dates of Sputnik and Vostok) betrayed their responsibilities. It was also possible to identify other members of the Academy of Sciences involved in the space industry, as indications of their specialties clarified their roles.

Thus in 1971 it was possible to identify the chief designer of cosmic engines as V. P. Glushko, the chief designer of gyroscopes as V. I. Kuznetsov (whose name appears in the history of gyroscopes), the chief designer of radio engineering systems as M. S. Ryazansky, the chief designer of guidance systems as B. E. Chertok (who was one of Korolev's deputies), and many others including A. Yu. Ishlinsky, S. S. Lavrov, V. P. Makeev, T. M. Eneev, N. A. Zheltukhin, N. S. Lidorenko, A. P. Vanichev, and V. M. Ievlev. The name of Vladimir Barmin also appeared not only as one of the inventors of the Katyusha rocket used by the Soviet Army in the Second World War but also as a specialist of refrigeration and compressors – the first evidence to identify him as the chief designer of the launch infrastructure at the Baykonur cosmodrome.

In 1971-1972 this data was checked against another important source – *Acts of the Communist Party Congress* of 1961, 1966 and 1971. These documents were widely published and contained a complete list of delegates, their profession and hometown. It soon became apparent that these "ingénieuro" ("technical workers") were primarily from the defence industries.

It was possible to identify the production centre of Dnepropetrovsk in the Ukraine (whose chief designer was Mikhail Yangel) as well as the 'South Machine

Building Plant' (Yu.MZ) situated there – which was officially tasked with producing tractors. Its director A.M. Makarov had signed the Yangel obituary, betraying products more strategic than tractors! Alexander Konopatov, a delegate from Voronezh in 1966, seemed to have superseded the engine designer Semyon Kosberg, who died in that city in 1965. V. P. Makeev was running a design office in Chelyabinsk which would prove to be the main builder of submarine-launched ballistic missiles.

Many other sources of information were used, including:

1. The short biographies published after a round of elections to the Academy of Sciences, published in Vestnik Akademii Nauk SSSR, and the lists of candidates published in *Izvestia*.
2. The Directory of the Academy of Sciences of Ukraine, which gave a complete view of the Ukrainian space research centres of Dnepropetrovsk and Kharkov.
3. The many obituaries published in the press.
4. Classified information probably published by mistake in open journals – two examples being the announcement by the journal *Priroda* of Viktor Makeev being awarded the 'Korolev medal' and the brief appearance of L. V. Smirnov (president of the state commission for piloted space flights) in a film released to the public.
5. Hundreds of books constituting, at that time, a unique collection in France.

On this basis we identified the real names and pseudonyms of V. P. Glushko (who wrote as G. V. Petrovich), B. E. Chertok (as B. E. Evseev), V. I. Kutsnetov (as V. I. Viktorov), Babakin's successor S. S. Kryukov (as S. S. Sokolov), V. P. Mishin (as M. P. Vasiliev), K. D. Bushuyev (as K. Davydov), V. P. Barmin (as B. Vladimirov), M. F. Reshetnev (as M. Fyodorov), and many others.

PUBLISHING THE RESULTS

Initially (1974-1982), the French space sleuths respected the anonymity of the Soviet writers, and used their nicknames. In 1974, for example, I quoted in a widely published study useful information that I found in articles by five academicians and corresponding members of the Academy of Sciences: S. P. Korolev, V. P. Glushko, M. V. Keldysh, B. E. Evseev (actually Chertok) and M. P. Vasiliev (Mishin, who was Korolev's first assistant).

In the early 1970s the *Cosmos Encyclopédie* first published a series of our articles reporting these Soviet space secrets. At the start of 1972 the *Encyclopédie* published a long article by myself and Albert Ducrocq describing the Lunokhod vehicle, including a description of the facility at Yevpatoria in the Crimea where a team drove it by remote control. The paper was a synthesis of dozens of articles in recent Soviet newspapers and magazines. At that time, I also published a very detailed description of the control panel of the Soyuz spacecraft, with all indicators at the disposal of the cosmonauts.

5.4: French sleuth Claude Wachtel in 1978.

Self-taught 'spy' skills

Day after day, I became an autodidact of open intelligence methods, having learned how to glean classified information from items published in the press. Human vanity and the thirst for honour allowed a lot of material to be published in the Soviet press – even if the person was only presented as a specialist of applied mechanics or as "a well-known scientist in the field of physico-technical problems of energetic" (which meant rocket engines). This skill led to my working at the Centre d'Exploitation du Renseignement Scientifique et Technique (CERST) during my military service in 1976.

In 1981, in the 'Twenty Years in Space' special issue of the magazine *Sciences et Avenir*, I described in detail the origins of the Vostok spacecraft of Gagarin. This included, for the first time, the names of the main assistant constructors to Korolev and Chertok, including the chief designer of life support systems, Grigori Voronin. I also first published the plans devised in 1964 by Korolev to use the Soyuz system for a manned circumlunar flight. This plan could have been realised using his own R-7 launcher without waiting for the Proton rocket (UR-500) that was being designed by the rival design bureau of Vladimir Chelomey.

In a full-page article in the daily newspaper *l'Alsace* in 1982 to celebrate the 25th anniversary of the cosmodrome at Baykonur, I first published the name of the chief constructor of the ground infrastructure, Vladimir Barmin, and described in detail his progress from the Kompressor refrigeration plant in the 1930s to creating the launching pad for Sputnik. The same year, I also published the first information on the head of the Ministry of General Machine Building, S. A. Afanasyev – about whom one chief constructor wryly pointed out, "The Americans have the NASA; as for us, we have Afanasyev." The article also mentioned the state commission for piloted flights and its two first presidents K. N. Rudnev and L. V. Smirnov, as well as the roles of Academician Valentin Glushko and the constructors N. A. Lobanov (parachutes), S. M. Alekseev (spacesuits) and G. I. Voronin (life support systems).

'Secrets' visit the West

In those years, Soviet chief designers were forbidden to travel to the West lest they seek asylum there. In 1971 Mikhail Tikhonravov, who in all probability was the real 'father' of Sputnik, attended the IAF congress in Brussels. No one knew that he had worked so closely with Sergei Korolev. Unfortunately, he died whilst attending the congress.

For some reason the Le Bourget Air Shows proved to be an exception to this rule, because each year the Soviet authorities allowed several deputy ministers and chief designers to attend. They visited under the cover of false affiliations of a minor high school or university in order to hide the fact that they led a large company or a design bureau. In this way Vladimir Barmin (chief designer of the launching infrastructure), S. S. Kryukov (chief designer at that time of the interplanetary stations "Venera" and "Mars") and N. D. Khokhlov (deputy of the Ministry of General Machine Building) came to Paris.

Clues to cosmonaut deaths

It was during the 1971 Le Bourget Air Show that the C2F hosted cosmonauts Pavel Popovich and Vitaly Sevastyanov at a restaurant near the Champs-Elysées on the same day that the crew of Soyuz 11 occupied the orbital station Salyut 1. This was my first real meeting with cosmonauts and it was a memorable moment. Three weeks later the Soyuz 11 crew lost their lives as they made their way back to Earth. Shortly afterwards in an article entitled 'Soyuz 11 – why the drama' in the journal *Cosmos-Information*, Albert Ducrocq revealed the cause of the accident as a valve which was to have automatically opened in the final phase of the parachute descent. He wrote that a pressure sensor on the valve had inadvertently been triggered at the instant of separating the descent capsule from the rest of the vehicle. In reality, Ducrocq wasn't guessing anything, as he had information from reliable sources (presumably a member of the Soviet Academy of Sciences), and he privately confided to me later that cosmonaut Patsayev unfastened his belt and vainly attempted to plug the leak. Ducrocq heard that the Soviets planned to launch Soyuz 12 with two cosmonauts in spacesuits to inspect the devices left aboard Salyut 1 by the Soyuz 11 crew, but the station was never reoccupied.

One year later, during a visit to an exhibition in Moscow, I was stunned to see in front of me a diagram with the much-talked-about valve. I came back with a drawing

5.5: Journalist Albert Ducrocq's contacts provided exclusives.

that was widely used in future publications of the Cosmos Club de France. At the Le Bourget Air Show in the spring of 1973 we held a lunch for Apollo-Soyuz cosmonauts Alexei Leonov, Valeri Kubasov, Anatoli Filipchenko and Aleksei Yeliseyev. What we didn't know at that time was that Leonov and Kubasov should have been the crew of the first N-1/L-3 lunar flight, and that, along with cosmonaut Pyotr Kolodin, they were to have flown Soyuz 11. But an anomaly on the EKG of Kubasov in the final pre-fight medical resulted in the backup crew of Dobrovolsky, Volkov and Patsayev flying in their place – and then losing their lives. The lunch was in the beautiful 'En plein ciel' restaurant of the Eiffel Tower. Kubasov explained me, in great detail, the functioning of the "Vulcain" welding system that he used on board Soyuz 6 in 1969. Yeliseyev, we discovered, spoke French, but cosmonauts were often discreet about their linguistic skills, probably having been instructed by Soviet counter-intelligence in order to prevent contacts with Western intelligence.

Last 'scoops'

At the BIS Soviet Forum in 1984, I gave a lecture detailing the organisation of the Soviet space industry, but this was published largely unnoticed in the *Journal of the British Interplanetary Society* the following year. I identified six major designers of the R-7 rocket: Korolev, Glushko, Pilyugin, Barmin, Kuznetsov and Ryazansky, as well as Korolev's closest associates: Okhapkin, Mishin, Raushenbakh, Ishlinsky, Bushuyev and Chertok, in addition to the designer of spacesuits (Alekseev), Voronin (life support systems), Lobanov (parachutes), and N. S. Stroev (for what in the West would be called avionics). Of the errors, the most notable was assigning all of the manned missions to the design bureau of Vladimir Chelomey whereas in reality the most important part of the programme was managed by the Energiya design bureau of Valentin Glushko.

The article clarified the role of the Ministry of General Machine Building, with its main leaders. It presented the space organisation of the Ukrainian Dnepropetrovsk centre, with key officials: Yangel, Utkin, Nikitin and Budnik. It identified other missile design bureaus headed by A. D. Nadiradze in Moscow, V. P. Makeev in Chelyabinsk, V. N. Chelomey near Moscow, and the designer of solid propellants B. P. Zhukov. It also identified the major consulting firms of satellites:

- The Office of lunar and planetary stations, then headed by V. M. Kovtunenko
- The Institute of electro-mechanics (A. G. Iosifian), which was responsible for meteorological satellites
- The firm of M. F. Reshetnev, which was responsible for navigation, geodesy and telecommunication satellites
- The design bureau of military observation satellites and R-7 launchers led by D. I. Kozlov in Kuybyshev.

French sleuths vindicated

Since 1989 a large part of the secrecy imposed on these organisations has been lifted, confirming many of the details published by the French sleuths. In an article that year in *Pravda*, Academician Mikhail Reshetnev recalled an animated discussion with an official responsible for security of travel who refused to specify that Reshetnev was a chief designer of one of the main space design bureau, insisting that his role was a secret. Reshetnev exhibited my article from the *Journal of the British Interplanetary Society* identifying him as the chief designer in charge of Soviet communication and navigation satellites as well as several other small military satellites. Unfortunately the official was insistent and Reshetnev visited Paris as a teacher from a university in Krasnoyarsk.

Over the years the documentation accumulated by the Cosmos Club de France enabled it to publish numerous 'scoops', including the astounding system utilised by the Luna 9 probe to make the first soft landing on the Moon in February 1966, which was protected inside an inflatable damping balloon that bounced several times on the lunar surface; a steering handbook of the Soyuz spacecraft published in 1980; and the first schematic of the Proton launcher's once secret RD-253 rocket engine. Whereas the existence of the Soviet manned lunar programme was only officially revealed in 1989, it had leaked out more than a decade earlier in a Soviet book unnoticed by the censor.

In 1977, General A. N. Ponomarev published a book about aircraft manufacturers in the USSR. The chapter dedicated to the aircraft engine designer Nikolai Kuznetsov mentioned that he had built rocket engines for a satellite launcher ("raketa nositel" in Russian). This immediately caught the attention of the French group as, at that time, all identified Soviet launchers were equipped with Glushko and Konopatov engines. The French observers weren't to imagine that the name of the main constructor of the engines for the lunar launcher had been revealed!

Seeing 'secrets' in person

In 1989 I was responsible for the 'Fobos letter' – published between February 1989 and December 1990 during the flight of the Fobos spacecraft to the larger of the two moons of the planet Mars. Since the mission carried several French experiments, the Cosmos Club de France addressed the Fobos letter to its members and to the main scientific laboratories concerned. One of the goals of this publication was to give to French scientists a better view of the Soviet industrial organisation in charge of the flight, as their activities were unknown of the majority of the scientific community. We described some episodes, largely unknown in the West and revealed to us by the Soviets, such as a planned sample return mission to Mars that was cancelled shortly after the departure of the chief designer S. S. Kryukov in 1977.

During the spring of 1991, I went to Russia and visited the Khrunichev factory in which the Proton launchers and orbital stations were built. I also had the privilege of visiting the hangar of the Moscow Aviation Institute where the relics of the Soviet manned lunar programme are preserved – most notably the vehicle of the type that would have enabled Leonov to land on the Moon. It was an unforgettable visit with Professor Alifanov. In this small cluttered room, we could barely squeeze between objects which used to be state secrets. Making a visit to the Exhibition of National Economic Achievements three months before the Soviet Union collapsed (not that anyone knew that was imminent, of course) I was surprised to find excellent models of the rocket engines that the aircraft designer Kuznetsov had put on the first stage of the super launcher N-1. It was 14 years after the revelation of Ponomarev's book.

After the end of the USSR, my role in the Cosmos Club de France would be less important, but I nevertheless presented a lecture in Paris in November 1991 about the new organisation of the Russian space industry. Soviet cosmonautics has lost its wall of secrecy, and with it has gone the opportunity to discover and publish totally new information. It was time for a new avenue of research work. I began a historical study of 'Star Wars' and published my first article about it in the French magazine *La Recherche* in March 1995. The *Journal of British Interplanetary Society* of March-April 2001 contained a detailed article on the subject entitled 'The lost Star Wars'.

From Le Bourget to Baykonur

by Christian Lardier

My passion for space began in March 1965, when I was thirteen years old, with the spacewalk of Alexei Leonov. My parents had recently obtained a black and white television, enabling me to watch the event a few hours after it occurred. In October, I joined the Cosmos Club de France and avidly followed the Moon Race.

SLEUTHING

In 1971 the deaths of designers Yangel, Babakin and Isaev were the subject of obituaries that provide extensive information on the space organisation in the Soviet Union. I then purchased from the Globe bookshop in Paris a copy of the book *Soviet Encyclopedia of World Astronautics* by V. P. Glushko (published by Mir in 1971). In October 1972 I went to Moscow for the first time, and found the book *Pioneers of Rocket Technology: Vetchinkin, Glushko, Korolev and Tikhonravov* (Nauka, 1972). While the first was in French, the second was in Russian and Claude Wachtel at the C2F advised me to learn Russian with a dictionary and grammar book. From then on I subscribed to *Pravda*, the journal *Aviatsii & Kosmonavtika*, and regularly ordered books on the Russian space programme.

In April 1973 I joined Air France, which enabled me to travel easily and my first ticket was for Moscow! In September of that year, I went to Amsterdam to attend my first IAF congress. There I met Maarten Houtman and Jacob Terwey – publishers of the journal *Spaceview*. They had been corresponding with Charles Sheldon, Marcia Smith, Charles Vick and James Oberg in the United States (this correspondence for 1973-1976 is now in my archive) and I began to work with them. I am still a good friend of Terwey. During the IAF congress, I asked Alexei Leonov (who was a very good painter) to make an exact drawing of the Voskhod 2 spacecraft, which was still secret at that time. He did, and I made Xerox copies at the press centre. That Monday night my bag with the drawing was stolen in Dam Square. Fortunately, Houtman and Terwey still had their copies!

That December I got married, and the wedding gift from my colleagues was a set of the 30-volume *Great Soviet Encyclopedia* containing biographies of all members of the Academy of Sciences. This allowed me to identify many officials of the space programme. For example, the identities of K. D. Bushuyev and V. P. Mishin were revealed in 1972. By then I had found in a small Russian bookshop in Paris (rue de l'Eperon) the book by Mishin, Lavrov and Appazov entitled *Ballistic of Long-Range Guided Missiles* (1966). My first article was published in the special issue No. 14 of *Science & Avenir* in 1974 on the results of Soviet Mars interplanetary missions.

Making contacts

In June 1975 I met V. S. Avdouievsky at the Le Bourget Air Show, and he gave me his address at the Moscow Aviation Institute. Two years later I would deliver to him a copy of our publication *Cosmos Encyclopédie*. It is now known that all of the Soviet designers were coming to Le Bourget using false identities. For example, Kerimov, Mishin, Lapygine, Budnik, Kourbatov and others all visited Paris, Bordeaux and Toulouse in 1965. And the crash of the Tu-144 counterpart to the Concorde at the Air Show in 1973 was witnessed by Baklanov, Balmont, Grichine, Kryukov and others.

At the Air Show I also met new friends from the GDR (East Germany): Karl-Heinz Eyermann and his photographer. I visited them in East Berlin in May 1981.

I couldn't go to Moscow for the Apollo-Soyuz mission in July 1975 but Maarten Houtman did and I later got to see the technical documentation that he collected. He had also received a NASA document on the Soyuz spacecraft in Houston, and this enabled me to create a complete description of the vehicle.

My third trip to Moscow was in April 1976 when I visited Victor Sokolsky at the Institute of Natural and Technics History (IIET AN SSSR). There I could get copies of bulletin *Iz Istorii Aviatsii & Kosmonavtika* and proceedings of the 'Tsiolkovsky Congress' held in Kaluga. I also spent a few days at the Lenin Library, where I was able to see the *Reaction Technics* and *Reaction Motion* bulletins published by RNII between 1935 and 1939.

In October 1976 I met Mitchell Sharpe of the Space & Rocket Center in Huntsville, Alabama. He later wrote me a letter requesting the names of the Soviet delegation that attended the British firing of a captured V2 rocket at Cuxhaven just after the war. I gave him the names of Generals Sokolov and Gaidoukov (the third, Tiulin, was unknown at that time) and he used the information in his book *The Rocket Team* (1979).

During my fourth trip to Moscow in April 1977 I visited the Institute of Medico-Biologic Problems (IMBP MZ) and met Oleg Gazenko and his team. I also visited the Institute of Space Research (IKI AN SSSR) to see the French Signe 3 satellite prior to its launch. The IIET suggested that I go to Riga to attend the 'Tsander Congress', but I didn't get this opportunity.

Becoming a space reporter

In 1978 I started to write regularly for *Espace & Civilisation* magazine, and I also

5.6: A rare 1982 sighting of Chief Designer Glushko.

occasionally contributed articles to *Sciences & Avenir*. I had two articles in a special issue in 1981 – 'The Creativity of Soviet Designers' and 'Baykonur Cosmodrome', the latter being written with the help of East German journalist Gerhard Kowalsky, who had attended the launch of Sigmund Jähn in August 1978.

Besides Russian material, there were other sources of information: the articles of Didier Laurent from Cedocar (Centre de Documentation de l'Armement de la DGA) published in *L'aéronautique & l'astronautique* from AAAF, the Pentagon's *Soviet Military Power*, and Nicholas Johnson's *Soviet Year in Space* reports. In 1981 *Soviet Military Power* reported that the Russians were developing a super-rocket (first flight expected in 1983) and that this would launch a large space station and laser weapons. Then it said they were developing: a super-rocket (Energiya!); a space shuttle similar to the NASA one (Buran!); a space station for 12 cosmonauts; a small shuttle (BOR mini-shuttle!); and a new medium-lift launcher (Zenit!). In March

1982, I went with Jacob Terwey to East Germany (Berlin and Leipzig) and Czechoslovakia (Prague). There I met a young engineer called Jan Kolar, who would later become the head of the 'Space Office' of the Czech Republic.

In June 1982 I went to Moscow for the first French-Soviet manned flight (PVH) and for the first time got to visit Soviet mission control in Kaliningrad (known from its acronym as 'Tsoup'), Star City, and the Tsiolkovsky museum in Kaluga. During my trip to Tsoup, I spotted Chief Designer Valentin Glushko. I tried to approach him to get an autograph but a security man prevented me and I was only able to take some pictures. I was very impressed by my encounter with this living legend! The French journalists permitted to travel to Baykonur to see the actual launch were: Serge Berg (AFP), Christian Sotty (ACP), Pierre Langereux (*Air & Cosmos*), Michel Forgit (France Inter), Michel Chevalet (TF-1), Georges Leclère (Antenne 2), Daniel Durandet (FR-3) and photographer Eric Préaut (Sygma). For the landing in Arkalyk in Kazakhstan, the press pack consisted of Jean-Paul Croizé (*Figaro*), René Pichelin (*Humanité*), Jean-François Augereau (*Le Monde*), Alain Raymond (AFP), Christian Sotty (ACP), Michel Chevalet (TF-1) and Eric Préaut (Sygma).

Early days of *glasnost*

In July 1988 I returned to Moscow, this time for the Fobos launch, but again was not allowed to travel to Baykonur. However, I met up with my very good Russian friend Leonid Journya of IMBP.

5.7: Christian Lardier with Buran in late 1988.

That September I was in Bangalore for the IAF congress, where I met Vladimir Prisniakov from Dnepropetrovsk (Ukraine). I told him that I knew about Ukrainian designers like Yangel, Budnik, Ivanov, Guerasiouta, Sergueiev and others. He was very surprised, and told me that I was a spy. I replied that I had found them all using 'open' information in the *Encyclopedia of the Ukrainian Academy of Sciences.* After that we became friends and I got to visit his house during a trip to Dnepropetrovsk in February 2002, and was able to see Vassili Budnik in a hospital just before he died.

In December 1988 I returned to Moscow for the French-Soviet 'Aragatz' manned mission, and this time I was allowed to visit Baykonur! I got to see Energiya-Buran and visited all the facilities for that programme including the integration hall, launch pad and the test bench from where the first Energiya was launched. My report of this visit was published in *Aviation* magazine. Exactly a year later I returned to Baykonur for the launch of the Granat satellite and was able to tour the Proton launch facilities.

Heavy schedule in the USSR

When in Moscow in February 1990 I visited the Energiya manufacturer, then paid my third visit to Baykonur to attend the launch of Soyuz-TM 9. In September 1990, I was in Moscow for the first aerospace exhibition at VDNK and that October I met for the first time Vasily Mishin at the IAF congress in Dresden. In January 1991, I visited the Khrunichev plant in Fili, the Monino Air Force museum which has the '105' aircraft built for the Spiral programme, and interviewed V. K. Novikov, general designer of the Myasishchev OKB. In March 1991 I attended the 50th anniversary celebration of the Institute of Test Flight (LII) and then in May visited NPO Lavotchkin in Khimki, where I met Oleg Ivanovsky at the institute's museum and was able to view the real Lunokhod 3 rover (whose launch was cancelled) and the much smaller 'Marsokhod' created for the Mars 2 and Mars 3 missions. During that visit I spent a full day with Oleg Gazenko at his dacha near Moscow.

In August 1991 I happened to be in Moscow at the time of the failed putsch! With the help of my friend Vladimir Vassiliev from the Ministry of Foreign Affairs (MID), I was accredited as an *Aviation* journalist for the following year – this was renewed annually until 1996. In December 1991 I visited TechnoMach, Lavotchkin (for the second time), Molniya (BOR and Buran) and Krasnaya Zvezda (nuclear generators). On 13 March 1992 I went to the Moscow Aviation Institute and saw for the first time the LK lunar lander, although unfortunately I was not allowed to take pictures.

On 14-18 March 1992, I returned for the fourth time to Baykonur for the launch of Klaus Flade on Soyuz-TM 14 and it was during his flight that I started writing for *Air & Cosmos* magazine. In April 1992 I went to the first engine exhibition (Dvigateli-1992) in Moscow and to an airplane exhibition in Kubinka. In July 1992 I returned for my fifth trip to Baykonur for the launch of Michel Tognini on Soyuz-TM 15. A month later I went to the second Mosaeroshow and I have been to all of them since.

My first book on Soviet space

Around that time I published my first book *L'astronautique Soviétique* (Armand Colin) using all the information that I collected in Russia during 1988-1992. With *glasnost*, most secrets were declassified. There had been a big change between the official 'TASS statements' story and reality. My book included all those changes and in his 2000 book *Challenge to Apollo* Asif Siddiqi said, "For those interested in the technical arcana of the Soviet space programme, [Lardier's] is the best book ever written on the subject. It uses much information declassified by the Soviets following 1988 and is incomparable in its breadth and ambition to any other book published on the subject in either English or Russian."

At the World Space Congress in Washington DC in October 1992 I met Vasily Mishin for the second time, as well as Charles Vick for the first time. The following month I attended the 'Rocket Propulsion Technologies Congress' in Moscow, then visited the Moscow Aviation Institute, the Keldysh Institute, NPO Energiya and NPO EnergoMach. Because I was now working permanently at *Air & Cosmos*, I wrote a lot of articles about Russia and Ukraine.

In February 1995 I was finally permitted to visit the once-mysterious Plesetsk cosmodrome, where I saw the launch of a Foton satellite. I was also allowed to look around Soyuz and Tsyklon pads. In August of that year I paid my first visit to NPO Machinostroenie in Reutov, where I interviewed G. A. Efremov, who had succeeded Chelomey in 1984. In the exhibition there I saw various naval ballistic missiles, the Polyot spacecraft, and an Almaz station and its film recovery capsules. Although I was not permitted to take pictures, by chance there was a large panel outside of the main building containing pictures of spacecraft and I was able to take one picture of an image showing the double launch of the VA capsule (Double Cosmos). After that I visited the Mojaïsk Military Academy of Saint Petersburg, and made many more trips to Baykonur. I visited NPO PM in Krasnoïarsk in 1998 and went to Samara three times. I also visited NPO Youjnoye in Dniepropetrovsk in 2002.

Preserving space history

After 1991 the Russians could not attend the history symposium of the IAF congress (V. N. Sokolsky had died in 2002), so I chose to help by giving my papers on Soviet space history. In 1996 I gave the paper 'Soviet designers when they were secrets' at the Beijing congress based upon my own research since 1972. My paper on Russian propulsion systems was presented in 1997 (ramjet), 1998 (solid propulsion), 1999 (liquid propulsion). These were extensively used by George P. Sutton in his *History of Liquid Propellant Rocket Engines* (AIAA, 2006). I also presented papers on the military space organisation (2000), the industrial space organisation (2005), and meteorological rockets (2008). I presented 'Soviet Rocket Engines from 1946 to 1991' at the AAAF Propulsion congress of Versailles in May 2002 and 'Electric and Nuclear Propulsion and Energetic in USSR' at the AAAF Propulsion congress of Heraklion in May 2008.

In 2010 I published my second book with co-author Stefan Barensky on the story of the Soyuz rocket (Edite). It has recently been translated into English. In 2011, I wrote (again with Stefan Barensky) another book detailing the history of the Proton rocket. For the last 20 years I have continued to collect all the books published by the Russians about their space programme – including those by the space institutes and factories. About 90 per cent of my library is in the Russian language. There are still lots of things to write about when it comes to Soviet-Russian space history and this will become my main preoccupation when I retire in 2013.

ORBITE 'SOVIET SPECIALS'

The Cosmos Club de France published 'special issues' of its bulletin *Orbite* with numerous Soviet 'scoops'. Examples include: January 1986 issue (secrets of the Proton); June 1986 ('Spécial Mir'); June 1987 ('Spécial Energiya'); October 1987 (30th anniversary of the space era); January 1989 ('Spécial Buran'); and September 1989 (the N-1/L-3 lunar programme).

Like the R-7 Semyorka rocket, which was a secret between 1957 and 1967, the Proton booster was an official secret between 1965 and 1985. In fact they were only declassified after the deaths of their designers – Korolev in 1966 and Chelomey in 1984. In a 1972 Soviet movie we saw the upper part of a Proton launching the first Salyut space station, but the first stage remained a mystery. Most Western analysts gave us their own interpretations, but these were all wrong. The sleuths imagined jettisonable boosters like an R-7, but no one guessed that it was a "barrel" of tanks similar to the Saturn I booster developed by NASA.

In their writings the Western sleuth's speculations on the thrust of engines, their number and the total thrust at lift-off were all incorrect. The standard model for the Proton was seven engines of 200t each with six boosters and a central core having a total thrust of 1400t: Charles Vick gave $6 \times 220t = 1320t$ in total plus ignition of the core at an altitude of 35-40 kilometres; Philip Clark gave $6 \times 240t = 1440t$ with a core of 300t thrust being ignited on the ground; Alan Bond and John Parfitt gave $6 \times 210t = 1260t$ with a core of 210t thrust ignited on the ground; and Ralph Gibbons gave $6 \times 250t = 1500t$ plus ignition of the core of 400t thrust at an altitude of 35-40 kilometres. Using the available data, Alain Souchier, Marcel Pouliquen, Jacques Villain, Claude Wachtel and I recalculate the exact thrust of the first stage engine as 150t at sea level. The characteristics of the RD-253 engine were published by V. I. Prichepa in the book *Isselodavania po istorii i teorii razvitia aviatsionnoï i raketno-kosmitcheskoï nauki i tekhniki* number III in 1984.

The first time we all saw the first stage was during the launch of the VeGa probes, which were shown on television in December 1984. Even then we had to wait for the new book *Kosmonavtika Encyclopedia* by Glushko to be published in 1985 to find out that it was a barrel of six RD-253 of 150t thrust each (900t in total). The C2F got a copy of this book through diplomatic channels. Claude Wachtel and I immediately published a special bulletin of *Orbite* in January 1986 and dispatched copies to many analysts around the world. After this January 1986 *Orbite* issue, the information from *Kosmonavtika Encyclopedia* was used in the article 'New insight into space

LE BULLETIN DU COSMOS CLUB DE FRANCE

JUIN 1987

I8 rue Saint-Benoît
75006   PARIS

HORS SÉRIE N°3

SPECIAL ENERGIA

SOMMAIRE

1

5.8: An *Orbite* special devoted to Energiya.

activity' by H. Pauw (*Spaceflight*, June 1986). English analysts writing in the article 'Proton Re-evaluated' in *Space* magazine (June-August 1986) claimed they were the first to find the correct characteristics of the Proton. It was not true! In 1986 another Soviet book, *Kosmonavtika SSSR*, published the first colour pictures of the Proton launcher. I got these pictures through my East German friends.

Soviet lunar project?

One of the 'hot topics' in the 1970s was the Soviet manned lunar programme. Early drawings of the N-1 rocket were made by Charles Vick – then known as the 'Lenin' or the G-1 booster – but they were incorrect. The first correct drawing of the secret booster was in an article by Craig Covault in *Aviation Week* on 16 June 1980. The drawing was signed by the unknown person "L. J. Herb". Vick revised his drawing and it was published in Kenneth Gatland's *Encyclopedia of Space Technology* of 1981 and also in the 1986 *Encyclopedia Universalis*.

The first real clue the Soviet Union provided was a drawing of the service tower published in the book *Kosmodrom* in 1977, but this was difficult to understand. The N-1/L-3 programme was declassified by the Russians 20 years after the Apollo 11 mission in an article by Sergei Leskov in *Izvestia* on 18 August 1989. The Cosmos Club de France published a special *Orbite* bulletin on the subject on 3 September and circulated it widely. The first public account of it was 'Why we didn't fly to the Moon', written by Mishin and published in December 1990. Prior to these new revelations, the dates of the N-1 launches were thought to have been June 1969, June 1971 and November 1972. The real dates were 21 February 1969, 3 July 1969, 27 July 1971 and 23 November 1972.

On my side, I must admit that I was sceptical because the Americans were doing a lot of misinformation to exaggerate the 'Soviet threat' and increase their own budget. I only recognised the lunar programme in 1989 when the Russians declassified it and then I saw the LK lunar lander during my first visit to the Moscow Aviation Institute in February 1992.

Searching for secret designers

Most Soviet designers signed with a pseudonym, and for better understanding we needed to know to whom they corresponded. As explained by Claude Wachtel above, this research first got underway in France back in 1965. I presented my own paper 'Soviet designers when they were secrets' at the 1996 IAC in Beijing. Asif Siddiqi published an article in *Spaceflight* (September 2008) on how Korolev was known in the West before his death in January 1966. Of course, he was recognised as a rocket pioneer before the Second World War (GIRD, RNII) and was present at 'Operation Backfire' in Cuxhaven in October 1945, but after that he had vanished into secrecy. On 27 September 1957 Korolev published an article in *Pravda* in his own name for the 100th anniversary of Tsiolkovsky's birth. And then in the Vestik of Academy of Sciences n°12, 1957, Blagonravov wrote an article that quoted the corresponding-members Korolev and Glushko. Tokaty-Tokaev, an engineer in the

Soviet air force who was in Germany in 1945 and defected to London in 1947, seems to have been the first to speak about them during a British Interplanetary Society lecture in 1961. Theodore Shabad, a Moscow correspondent for the *New York Times*, wrote articles quoting them in 1963 and 1965. Finally, the Aerospace Information Division of the Library of Congress identified Korolev as the mysterious Chief Designer in a study entitled *Top Personalities in Soviet Space Program* published in May 1964.

Just like the designers, the identity of the president of the state commission for space launches was secret. We knew of K. N. Rudnev for the flight of Gagarin in 1961, and L. V. Smirnov for the Vostok 3 and Vostok 4 joint flights in 1962. The first articles on this subject appeared in *Zemlya i Vselenaya* (n°5/88, You. A. Skopinsky), *Aviatsia i Kosmonavtiki* (n°10 & 11/1988, B. A. Pokrovsky), and *Zemlya i Vselenaya* (n°5/90, A. A. Maksimov) where we discovered the roles of V. M. Riabikov, G. A. Tiouline, K. A. Kerimov and others. In 1978, with the Intercosmos manned flights, members of the press were invited to 'their' launches in Baykonur. A friend of mine, Gerhard Kowalski, took a picture of the state commission for the DDR cosmonaut Sigmund Jähn. When I was in Berlin he gave me a copy and I spotted Kerimov and Glushko with the crew. I saw the state commission myself when I was in Tsoup for the French PVH mission of June 1982.

Unknown cosmonauts

Right from the start of the Space Age there have been rumours of dead cosmonauts. In March 1965 the newspaper *Le Figaro* reported that the Judica-Cordiglia brothers, who were radio amateurs based in Turino, Italy, had recorded a total of fourteen dead cosmonauts between 1960 and 1964. In May, a study by Julius Epstein of Stanford University in America listed a dozen dead cosmonauts. One of them, Pyotr Dolgov, did in fact lose his life during a flight of the Volga stratostat balloon in November 1962. Another rumour said that the test pilot Vladimir Ilyushin, a son of the aircraft designer, had taken part in a failed space flight before Gagarin. In *Air & Cosmos* in April 1965 Albert Ducrocq wrote: "This information is dismal in all respects. In the public mind they cast doubt."

There were cosmonaut groups, and we were seeking information on the 'missing' men who did not fly. For example, only 12 of the first group of 20 cosmonauts went into space. The eight missing cosmonauts were disclosed in *Izvestia* on April 1986 to mark the 25th anniversary of Gagarin's flight. For the first time we learned about the death of Bondarenko during a ground simulation and the fate of Nelyubov, a support cosmonaut for Gagarin, who committed suicide in 1966.

The female group of 1962 included five women, but only one was known to us. Tereshkova referred to the others as Vera, Irina, Tatiana and Janna. In June 1988 the names of the four girls were finally given as Valentina Ponomareva, Irina Solovyeva, Tatiana Kouznetsova-Pitskhelaouri and Janna Erkina-Sergueytchik.

During an event that I organised in Orly in October 1987 for the 30th anniversary of Sputnik, I met Feoktistov and asked about backups for the Voskhod 1 flight – he immediately answered 'Katys' and 'Sorokin'. This was the first time that I had heard

5.9: Lardier comes face-to-face with cosmonaut Feoktistov in 1987.

those names. More information was given by Sergei Shamsutdinov and Igor Marinin in 1993 when they published 'Flights that didn't exist' in *Aviatsia I Kosmonavtika*. Since then everything has been declassified about the cosmonaut group. The Western specialists on this topic were James Oberg, Gordon Hooper, Rex Hall, Marc Hillyer, Bert Vis and, in France, Michel Clarisse.

The secrets of Almaz

The Salyut space station programme included two different models: the civilian one made by Mishin's bureau and the military one made by Chelomey's bureau, although they had a common heritage. The second one was secret and no pictures or drawings were published of the station at that time. The first drawings were published in December 1991 in Igor Afanaseyev's book *Unknown Spacecraft*. I used his new information in my own book in 1992. Then there were articles in *Krylia Rodiny* (1992, V. A. Poliatchenko), *Aviatsia I Kosmonavtika* (1993, Poliatchenko and Toumanov), *Spaceflight* (1994, Neville Kidger), *JBIS* (1994, V. M. Petrakov) and a booklet by Olaf Przybilski in 1994. The full story was finally written by Asif Siddiqi for *JBIS* in 2001-2002.

Mystery of the 'Double Cosmos' flights

Another subject of great interest concerned the so-called 'Double Cosmos' launches of December 1976 (with Cosmos 881 and 882), August 1977 (failure), March 1978 (Cosmos 997 and 998) and May 1979 (Cosmos 1100 and 1101). They only made one orbit before re-entry and Western analyst thought they were shuttle tests. This resulted in articles such as 'Soviet build reusable shuttle' (Craig Covault, *Aviation Week & Space Technology,* March 1978) and 'Soviet re-entry tests: a winged vehicle' by Trevor Williams (*Spaceflight,* May 1980) but it was not until the declassification of Chelomey's Almaz military space station programme that we found out they were tests of pairs of the 'TKS' manned capsule. (In the new era of openness, in December 1993 one of these capsules was sold at auction in America!)

CONCLUSIONS

In conclusion, I will say that some of the Soviet secrets were well-kept. The Proton, N-1/L-3, Almaz, the Double Cosmos flights and the 'missing cosmonauts' were not exposed by Western analysts before the Russians declassified them. Sometimes the analysts were totally wrong. It was not an easy job!

6

Orbital elements of surprise

by Phillip Clark

I was born in Bradford, West Yorkshire in 1950 and by the age of ten I had developed an interest, like many children, in prehistoric animals and astronomy. In 1962 the first American piloted orbital flights began and my teacher at Wibsey Junior School, Mr Slater, brought a radio into the classroom so that we could listen to these flights. We also had Earth globes and maps to enable us to see where he was overflying, and those afternoons were given over to following the flights of John Glenn and Scott Carpenter.

ASTRONAUT SCHOOLDAYS

Thankfully I lived about five minutes away from the school, so I could quickly get home and carry on listening to the coverage. Of course, I can also remember the warm, clear-skied Wednesday of 12 April 1961 when Yuri Gagarin became the first man in space – although I knew nothing about it until the evening's *Telegraph and Argus* newspaper arrived with a 'MAN IN SPACE' banner headline. That is the first space event that I can recall with any certainty.

After passing the "Eleven-Plus" examination, in September 1962 I was accepted for Grammar School, which meant that for the first time I had to take a bus journey on my own to the other side of Bradford's centre. Although I was decidedly average in maths, by the time I was fifteen years old I had found that lots of things to do with astronomy were simply mathematics.

I am able to remember the major events from 1962 onwards and a new era opened with the launch of the United States' Early Bird in April 1965: renamed INTELSAT 1, this was the first commercial telecommunications satellite in geosynchronous orbit and because it was positioned over the Atlantic it enabled Europe to receive live transmissions from the United States. We had a school half-term holiday starting on Thursday, 3 June 1965, and for the first time ever we were able to watch the launch of a crew into orbit, as Gemini 4 lifted off to make America's first spacewalk. I was

6.1: Phil Clark pictured in 2006.

able to follow the Gemini programme in detail, with quite a few events happening to fall on weekends or school holidays. Gemini 5 was during our long summer holiday, the cancelled attempt to dock Gemini 6 with an Agena target vehicle was during our half-term break, and the launch and landing of Gemini 7 were both on a Saturday and covered live. And communications satellites meant that we quickly got the pictures of the Gemini 7/Gemini 6A rendezvous in orbit over here.

In 1966 only Gemini 9A was convenient to watch on TV, coming during a half-term break. So did the landing of Surveyor 1 on the Moon on 2 June and for the first time British television used the caption "Live from the Moon" as the pictures came over live. At this point, I should mention that morning television on the BBC was an extreme rarity.

Of course, 1967 started with the shock of the Apollo 1 fire, killing Virgil Grissom, Edward White and Roger Chaffee on 27 January. Then on 24 April the ill-prepared Soyuz 1 flight ended with the descent module crashing under a fouled parachute and Vladimir Komarov being the first person to die during an actual space mission.

Maths for work and pleasure

With the O-Levels coming along in June 1967, the term between Christmas 1966 and Easter 1967 was devoted to learning calculus and suddenly for me everything seemed to click into place with maths. Due to family reasons, university wasn't an option and I left school and got a job.

Once I started work on 31 July 1967 at the building society I was able to begin to buy all of the books that I wanted, and with Apollo approaching there were plenty of books around. October 1967 was the tenth anniversary of the first satellite, Sputnik. And thanks to the local media I was able to follow Venera 4 parachuting through the atmosphere of Venus, with data being returned from inside the atmosphere of another planet for the first time. Then we had the unmanned Cosmos 186 and Cosmos 188 docking in orbit, as had been intended for Soyuz 1 and the cancelled Soyuz 2.

It was expected that I would do building society-related studies, but after finding them boring I decided that in the summer of 1968 I would register for an A-Level in mathematics: the local college only did evening courses for a combined pure and applied maths, which was half and half. This was a two-year course and after passing that I decided to enrol for a pure maths A-Level and study at home. I passed it a year later.

Holiday on the Moon

With 1969 my interest in spaceflight was becoming more and more serious, and I was starting to become interested in using my maths to understand the technicalities of the subject. As soon as the Apollo 11 crew and launch window were announced with the decision that this would be the first attempt to land a crew on the Moon, I booked the two weeks starting 14 July off work as my annual leave.

Like many 'space buffs' at that time, I spent the whole of the Apollo 11 mission sitting in front of the television watching everything live. The BBC's team headed by James Burke and Patrick Moore provided far better coverage than the commercial ITV station. In the UK, Eagle's landing on the Moon came at 20:17:39 GMT on 20 July and Neil Armstrong stepping onto the Moon came at 02:56 GMT on the next morning. Ever since then I have referred to 20 July as being "MoonDay" (despite the landing occurring on a SunDay) and 21 July as "Armstrong Day", raising a glass in memory on both days: the time zones mean that Americans only have the one day to celebrate both of these events.

After Apollo 11 it became normal for me to book time off work to coincide with the periods that crews would actually be on the lunar surface. This meant that I was able to follow most of the unfolding Apollo 13 story live, and that sort of made the Tom Hanks film redundant. The sole exception was Apollo 14, which coincided with the switch from our old "pounds-shillings-pence" currency to the decimal currency. There was no opportunity to take time off work to follow the lunar landing and the first day's surface activities, but I did get a telephone call from my mother who left the telephone by the television to enable me to listen to the final few minutes of the descent to the lunar surface. I did get some very strange looks from colleagues, but I was already looked upon as a mad scientist! Or maybe just mad!

Attention switches to Soviets

Apollo 11 marked the start of a new quest for me, one which ended up taking over most of the rest of my intellectual life. Until July 1969 the media had been saying how far ahead the Soviets were in the Space Race, yet there were the Americans on the Moon and no sign of a Soviet response. Okay, they had flown two unmanned craft around the Moon and apparently recovered them (only later did the Russians acknowledge that the second of these missions, Zond 6, had crash-landed) but there was no other obvious sign that they were trying to challenge the Americans in landing a crew on the Moon. What had happened?

Soviets in Space by William Shelton was published in Britain in the summer of 1969 and was plugged as the first Western book to tell the story of the Soviet space programme. My ordered copy actually arrived on 11 July, which was the day that I finished work to start my Apollo 11 summer holiday. This was my introduction to someone attempting to make sense of the Soviet space programme – although like many books of that period it concentrated on the piloted, lunar and interplanetary missions.

In December 1968, Geoffrey Perry, the head of the physics department at the Kettering Grammar School, had the first in a series of articles published in *Flight International* which reviewed the first 250 Soviet satellites in the Cosmos series. Over the months the magazine ran more of his articles.

I was curious about one odd-ball flight that wasn't properly explained – Cosmos 159 was launched into an orbit which reached out to around 60,000 kilometres and the orbital inclination suggested it was related to either the manned programme or to the deep space programme. At that time the Soviets were known to use the 'Cosmos' name to cover up launch failures in the lunar and planetary programmes. I reckoned Cosmos 159 might be a lunar probe which had got out of low Earth orbit and been stranded in a highly elliptical orbit by the premature shutdown of its fourth stage. At launch the Moon was close to its 'first quarter' phase, which it was roughly for the Luna missions in 1965-1966 that attempted to land on or orbit around the Moon.

Contacting "Mr. Perry"

I had no idea that during the summer months the Perry family migrated to Bude in north Cornwall, and it would be some weeks before I got a reply to my July 1970 letter to him. Surprisingly he took me seriously and explained that while the Moon was of the appropriate phase for a lunar mission when Cosmos 159 was launched, there was a question that I had not asked: "Where was the Moon at the time of the launch?" In those days I did not know to ask the question. I had heard of something called the Greenwich Hour Angle (GHA) in terms of the positions of stars relative to an observer. He pointed out that the Soviet lunar missions that started in 1963 had been launched when the lunar GHA was about 240 degrees; when Cosmos 159 was launched the GHA was about 60 degrees, so it was launched *away* from the Moon. He also dropped in the information that the Soviet Zond 4 probe was also launched away from the Moon, whereas Zond 5 and Zond 6 flew around the Moon.

That was the start of a long correspondence and friendship which continued until Geoff Perry died in January 2000. To me he was always "Mr. Perry", and I always considered him to be my mentor in terms of studying the Soviet space programmes. In those early days, most of the public analysis came from "The Kettering Group", which he had unintentionally founded. Being the head of the physics department he had decided to use spaceflight as a way of engaging pupils' interest in science. They learned about the scientific method, how to collect and analyse data and to prepare it for publication. When the papers appeared in print, the pupils' names were listed as co-authors. In 1966 the Soviet Union introduced its third launch site at Plesetsk, and by plotting the launch tracks the Kettering Group was able to deduce the location of this new base. Their analyses enabled researchers in America to talk about the new base, even though its location was still officially classified in the United States and the Soviets didn't officially announce its existence until the early 1980s. Mr. Perry said that the media frenzy at Christmas 1966 when the story broke of "schoolboys discovering a new, secret Soviet launch site" drove him to give up alcohol for life!

Mr. Perry also pointed me at two other sources of detailed information. One was the *Tables of Earth Satellites* issued by the Royal Aircraft Establishment (RAE, as it then was) in Farnborough, which was by far the most detailed source of orbital data that was available in the 'open' domain. I had first come across this via the British Interplanetary Society and had managed to get the annual issues from Farnborough. In September 1970 Dr Desmond King-Hele, who was head of the space department at RAE, added my name to the more restricted list for the monthly supplements and these provided me with details of new launches in a much more timely manner.

Geoffrey Perry was the most influential analyst of the Soviet space programme, because most of the people who subsequently did their own analysis (published or otherwise) were influenced by either knowing him or reading his publications. It is difficult to imagine how the 'hobbyist' interest in the Soviet space programme could have progressed without his influence.

Dr Sheldon's 'Green Reports'

The second source to which Mr. Perry pointed me was the *Soviet Space Programs*, a series prepared by the Congressional Research Service for the Committee on Science, Space and Technology, US House of Representatives. Starting with the 1966-1970 volume, these were prepared by Dr Charles S. Sheldon II and they had green covers, so they were referred to in the community of Soviet space watchers as 'The Green Reports' or 'The Sheldon Reports', even though Dr Sheldon had only taken overall control of them with the 1966-1970 volume. After my success in getting the RAE *Tables*, I contacted Dr Sheldon and he put me on the list to receive the volumes as they appeared.

I corresponded irregularly with Dr Sheldon, because his work meant that he was never a prolific letter writer, but I learned to 'read between the lines'. If he ignored a suggestion or piece of analysis then I knew that it was wrong. On the other hand, if he commented that it was "interesting" then it was accurate but he could not tell me that.

Dr Sheldon was able to reveal to the world what the launch vehicle that we now know as the N-1 actually looked like in the first volume of the 1971-1975 report. He showed the Type-G launch vehicle (as he called it) to have an overall tapering shape. Everyone presumed this to be speculation, as he took the shape of the Soyuz launch vehicle and scaled it up to show the approximate size of the then-secret N-1/Type-G. But with the advantage of hindsight, the outline of that drawing matches the now-declassified design of the N-1, including even the payload shroud and escape rocket tower of the Soyuz launch vehicle.

First sleuthing success

In 1970 I mentioned something to Mr. Perry which struck a chord with him since I was doing some numerical analysis. I had noticed that the later in the day a Soviet manned space mission was launched the longer the orbital lifetime would be. At the time I was just working with Soyuz 1, Soyuz 2/Soyuz 3, Soyuz 4/Soyuz 5, Soyuz 6/Soyuz 7/Soyuz 8 and (when the relationship was first noted) Cosmos 110 (which was not manned but had carried two dogs in orbit for around three weeks in 1966). The relationship wasn't perfect because there were some offsets due to rendezvous considerations (with Soyuz 2 being offset from Soyuz 3, for example). However, it suggested that the Soyuz 9 mission launched on 1 June 1970 would last for about 18 days. In fact the two cosmonauts were in orbit for 17.7 days, which wasn't too bad!

Mr. Perry suggested that I publish this analysis through the British Interplanetary Society but they were not interested – even though Charles Sheldon mentioned the relationship in his 1971 "Supplement" to the 1966-1970 Green Report. Eventually it appeared in a small-circulation publication about manned spaceflight by Eddie Pugh and then a different write-up appeared in the December 1973 issue of the *Journal of the British Astronomical Association*.

In 1972 Mr. Perry put me in touch with Ralph F. Gibbons, based in Chesterfield. We were interested in trying to make sense of all of the Cosmos missions. Two things

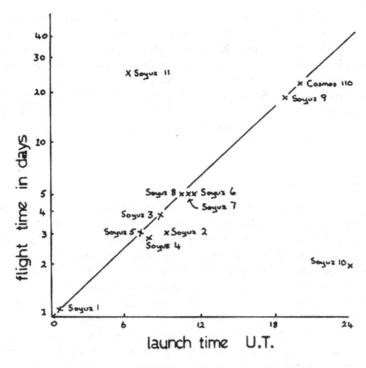

6.2: An attempt by Clark to predict Soviet flight durations.

which Ralph could do that I decidedly could *not* do were read the Russian language and make technical drawings of spacecraft. When he sent to me his notes about the different satellite groups (at that time photocopiers were not readily available), I was able to see how he was grouping the satellites compared with my own thoughts. We would happily bounce thoughts off each other as we tried to fathom out what each type of spacecraft was actually doing.

EARLY COMPUTER SKILLS

In 1974 I started to study with the Open University (OU) for a degree in maths (by coincidence, so did Ralph, but we did not mention this to each other until we started studying) and the eventual access to their computer system opened up a whole new way of calculating for me. My OU studies finished with an honours degree in maths and computing in 1982. I had started out using the paper tables for logarithms and other functions. That did not change when I got my first calculator in 1970, as it had only the four basic arithmetical functions and no memory. With the OU, however, I learned the BASIC programming language and sat down to write a programme based upon an RAE study of patched-conic lunar trajectories which would enable me to do some kind of calculations of lunar missions. I was also able to get a programme onto the OU computer that would calculate trans-planetary trajectories, and this resulted in a two-part paper being published by the British Interplanetary Society in their monthly *Spaceflight* in 1975 and 1976.

I was also looking at the Soviet launch vehicles, although my thoughts at that time were influenced by the work of Charles P. Vick on the Proton and 'Lenin' (what we now know as the N-1 launcher, intended to rival the Saturn V) in *Spaceflight* in December 1973 and March 1974.

Predicting missions accurately

I kept returning to the manned space programme, looking at landing times versus the date of landing. This showed that apart from some flights which were believed by Western analysts to have ended ahead of schedule (Soyuz 10, Soyuz 11, Soyuz 15) the landing times clearly varied with the seasons (for simplicity I was using GMT). This made sense because the lighting conditions in orbit had to be "right" for the time of retrofire, which in turn determined the time of landing.

In 1977 I had the idea of adding another parameter to this graph, one which meant a different graph for each year. Starting in 1971, most Soviet piloted missions flew to a space station and the nominal landing times were governed by a cycle of about 60 days between passes over the landing site at the right time of day for a landing. Once this was put together, not only were the presumed failed 'early return' missions more obvious, it was also possible to see the approximately week-long landing 'windows' these missions would probably have used if they had flown their planned durations.

This graphical method came into its own with the launch of Salyut 6 in September 1977. The Soyuz 25 crew launched on 9 October into an orbit that was perfect for a rendezvous with the space station, but they were unable to dock and returned to Earth much earlier than intended. As this was the 20th anniversary of the launch of the first Sputnik, presumably the Soviets had hoped that the mission to the new space station would help with the celebrations!

Armed with the knowledge that the nominal landing conditions occur every two months, it was apparent that another mission with the same duration as intended for Soyuz 25 could be launched about a third of the way through December. And indeed on 10 December Soyuz 26 was launched. The graphical method showed that there would be a landing window in mid-January 1978, so I wondered whether the Soyuz 26 crew would return then and be replaced by a Soyuz 27 crew. The reality was that Soyuz 27 was launched on 10 January and its crew returned aboard Soyuz 26 on 16 January during the landing window. But the Soyuz 26 crew remained in orbit until the next landing window in mid-March, when they returned to Earth after a record-breaking flight of more than 96 days.

I decided to see how the graphical method would predict other activity in 1978. Without knowing how long a flight was going to be it was impossible to work back to a launch date, but once a launch had occurred the landing opportunities were readily predicted. Soyuz 29 was launched on 15 June and it was clear that a landing window was approaching. Soyuz 28 had been the first of what would be a series of eight-day missions to Salyut 6 with a "guest cosmonaut" on board, and the launch of a Polish guest cosmonaut was widely expected. There was an opportunity for this eight-day mission to be launched in the last few days of June and recovered in early July. The launch came on 27 June and the landing was on 5 July.

The Soyuz 29 crew remained in orbit and the question was how long they would stay aboard Salyut 6. I had postal discussions with a few people but no one believed there would be a flight longer than about 100 days. In order to set a new record the Soyuz 29 crew would have to remain in orbit for at least 10 per cent longer than the previous record – 106 days. This duration did *not* fit with any landing opportunity. If the landing opportunities alone were considered, then the flight would last for either about 80 days or about 140 days. I was telling anyone who would listen that the crew would return in early November after a mission lasting about 140 days. They landed on 2 November after 139 days in orbit.

After this, I wrote up the analysis. It was published by *Spaceflight* in June 1979 (by then I had moved from Bradford to London as part of my work). The published paper included the predicted landing windows to be used in 1979 and 1980: if I had carried on until 1981 then I could have accurately predicted all the landing windows for Salyut 6 before it was abandoned in May of that year.

The summer of 1979 was my first chance to see a Soviet spacecraft. There was a major economics exhibition in Earl's Court in London, and it had a section dealing with the space programme. I was a regular visitor there, making the most of a chance to inspect Soviet spacecraft (mockups of course) and to talk to the Russians on duty to answer questions from the public. They soon got to recognise me – knowing that they would be on the receiving end of detailed questions which virtually no one else would even dream of asking. At one point they agreed to let me get into the sandpit with the Luna 16 sample-return spacecraft mockup and the Lunokhod lunar rover. While I was underneath Luna 16, looking at the engine arrangement I am sure that the Lunokhod being driven towards me was purely coincidental!

Solving Cosmos mysteries

For me, the really serious work in trying to understand the Soviet space programme was not looking at the manned programme – everyone seemed to be doing that. For me, the major interest was the all-encompassing 'Cosmos' programme. Launches to orbit had started in 1962 and by the end of the 1980s there had been more than 2,000 Cosmos satellites launched into orbit (plus plenty of launch failures which we now know about). It became a 'sport' to try and classify the Cosmos satellites in order to figure out what each satellite group was being used for. Some launches were failed unmanned lunar and planetary missions, while others were unmanned tests of future piloted spacecraft. These were the minority of launches, however.

In his *Soviet Space Programs* volumes Dr Sheldon had described the basic types of launch vehicles which were operating, linking the satellite launchers in most cases to modifications of missiles that had been put on public display, normally during the

6.3: Alongside a 'Cosmos' satellite of the Bion type, a modification of the Vostok.

May and November military parades in Moscow. He had also indicated from which site you could expect to launch into certain orbital inclinations.

The TASS launch announcements would allow a preliminary classification of the satellites. For example, there were flights from the Kapustin Yar launch site (at that time not officially acknowledged to exist) into orbits inclined at 48-49 degrees to the equator – these were small satellites and used the smallest launch vehicles. Launches from Baykonur (then still called Tyuratam in the West) were either at inclinations of 51-52 degrees or close to 65 degrees – most were of satellites which appeared to be variants of the Vostok manned spacecraft but there were also the lunar and planetary probe failures (identifiable because they were launched at the ideal opportunities for such missions). The Vostok-like satellites were in lower altitude orbits than we saw from Kapustin Yar and they disappeared from orbit after a week or two because they were being recovered. When Plesetsk became active we had a mixture of the small satellites in 71-degree and 82-degree orbits (a little later, 74 degrees and 83 degrees) and the recoverable satellites flying at 65-66 degrees and 73 degrees.

Thus, by looking at the announced orbits of satellites and knowing the probable launch vehicle and launch site, it was possible to easily identify the major series of satellites sharing the Cosmos banner. By far the largest group were the recoverable ones, with more than 30 launches per year until the demise of the Soviet Union and the general decline of its space programme.

Dr Sheldon had said that when looking at the Soviet space programme the most important thing one can have is hindsight, so for the experienced analyst something that was new would be readily identified as such, possibly also identifying probable precursor missions. His analyses of the satellite groups, which he published in the *Soviet Space Programs* volumes, were excellent pointers to the probable missions being undertaken by the different satellite groups. Of course, with the demise of the Soviet Union, the individual design bureaus are publishing historical information on their satellite programmes and so questions remain about only a very few satellites. What has been particularly pleasing is that the majority of the post-Soviet revelations have confirmed the deductions of the Western 'space sleuths'.

THE SLEUTHS MEET AT LAST

I had got to know Anthony Kenden, a software engineer based in the London area, very well and he had an interest in the United States military space programme – as well as more than a passing interest in the Soviet space programme. Anthony was a member of the BIS's programme committee and he had come up with the idea of a series of evening meetings, each on a different topic at a reasonable technical level. The first of these meetings, held in January 1980, was to be devoted entirely to the Soviet space programme.

During the 1970s the BIS magazine *Spaceflight* had become *the* publication for analyses of the Soviet space programme, although some writers used other outlets as well. It was routine for articles to appear in that magazine by Geoff Perry (primarily about with Cosmos photo-reconnaissance satellites), myself, Ralph Gibbons, James Oberg, and Charles Vick. In addition, the 'letters to the editor' section was regularly dominated by discussions of the Soviet space programme, bringing in names like Rex Hall, Michael Richardson, Nicholas Porter, Brian Harvey, Bert Vis, Bart Hendrickx and many others.

This first Technical Forum meeting was therefore a meeting of many people who knew each other by reputation but had never met one another before. Ralph Gibbons travelled from Chesterfield, Michael Richardson from Bournemouth, Geoffrey Perry from Kettering, Brian Harvey from Dublin, and Nicholas Johnson (who was at that time awaiting publication of two books about the Soviet space programme) from the United States. Anthony Kenden was the chair of the meeting, with him, myself and Rex Hall, whom I did not know at that time, all living in and around London.

The meeting was a great success and would subsequently be expanded to include Friday evening and the next Saturday morning and afternoon. This was *the* meeting place for people interested in analysing the Soviet space programme, and there was always a visitor from the Soviet Embassy to take note of what was being said.

Secrets in a communist bookshop

As a result of the meeting, Anthony kept in touch with Rex Hall and he arranged for us to meet up with Rex and his wife Lynn (who suffered 'space talk' with honours over the years). We all routinely visited Colletts Russian Book Store on the Charing Cross Road. That is where we normally headed on different days at different times to find copies of *Aviatsiya I Kosmonavtika* and any other Soviet magazines or books on spaceflight in the hope of finding secrets in them.

There was a public house behind Colletts, and that became the meeting ground for Anthony, Rex, Lynn and I, and we would spend whole evenings trying to thrash out what the Soviets were up to. Given what we now know, it is a fortunate there are no tape recordings of those meetings!

The meetings at the British Interplanetary Society have continued as an annual event, usually on the first Saturday of June, and they have forged many friendships. All was not taken seriously though, as I discovered to my cost. At the 1981 meeting I confidently predicted that with the introduction of the Soyuz-T spacecraft variant, the Soviets would cease sending the unmanned Progress cargo freighters to their Salyut space stations. I would be regularly reminded of this prediction as Progress craft, in different variants, continued to fly during the 1980s and 1990s. In fact, they are still in use as the regular cargo freighters to supply the International Space Station.

'Two-line elements' in the post

A major change in the data which was available to me came in August 1980. I had heard of the Two-Line Orbital Elements (TLEs) which Geoff Perry was using for detailed analysis, and these also served as the basis of the RAE's *Tables of Earth Satellites*. So I wrote off to try and get on the mailing list for these, not being sure what I was letting myself in for.

'Two-Line Elements' are measurements of the orbits of everything that was being tracked in orbit and was included in the official catalogues like the *Satellite Situation Report* issued quarterly (at that time) by the Orbit Information Group (OIG) within the NASA Goddard Space Flight Center. They allowed you to calculate reasonably accurately where a satellite was in its orbit. Without warning, my first mailing of TLEs arrived in August 1980. They were then issued three times per week (Monday, Wednesday and Friday) and the people who use online ones these days don't know how lucky they are. The pages were approximately A4 in size and covered single-sided. They were rarely in any sort of order and they could be oriented at any angle. You got what happened to be photocopied at the Goddard Space Flight Center and how it happened to be photocopied. To find the satellites you were interested in you generally had to go through several hundred of these sheets.

When I started to receive them, I developed programs for the Texas Instruments TI-59 calculator (by far the best calculator I have ever come across) which could be stored on magnetic cards. Many of the algorithms from those programs are still in use, albeit completely rewritten to run in the MS-DOS environment of a computer.

The TLEs started to be supplied purely in electronic format by OIG in mid-1994 and with suitable archiving routines (the originals were written by Peter Hunter in Australia) it was possible to do bulk calculations in seconds that would have taken many hours to process with a calculator.

In early 2005 it was planned to close down the original OIG computer system and replace it with a new system called Space-Track. The last data issued by OIG was on 4

February, and then its computer system suffered a major failure from which it could not recover. The new Space-Track system was already up and running, although OIG had been expected to continue for another month or so. Since then, the Space-Track website has been the primary source of TLEs through to the present day.

A comment must be made about the classified satellites. In addition to the orbital data for American military satellites being withdrawn, we no longer receive data for the Japanese, Israeli, French, German and Italian reconnaissance satellites – i.e. the really interesting satellites. But this does not mean that data isn't available. There is a worldwide unofficial network of satellite observers who regard spotting the secret satellites as a 'sport'. Their observations are collated to enable unofficial TLEs to be generated and made available on the Internet for anyone to use.

The long road to my own book

The BIS Technical Forum meeting in June 1981 was momentous for one paper that was presented. Claude Wachtel had come over from France and proceeded to give a history of the Vostok and Soyuz spacecraft, including variants that we had never seen before. His paper was a major shock, because no one knew that such information was available. He had a copy of a Russian-language book which we had never heard of. In English its title was *The Creative Legacy of Sergei P. Korolev.*

When I was eventually able to get my own copy of the book, Ralph Gibbons did the appropriate translations. These were mulled over, calculations were done, and I suggested to Ralph that we do two joint papers which would incorporate Claude's data into what we already knew or suspected. We agreed that Ralph would write a paper (with some input from me) dealing with Vostok and Voskhod, while I would do a paper (with some input from him) dealing with the Soyuz spacecraft. The two papers were presented at the Technical Forum meeting in June 1982, but owing to illness Ralph was unable to attend and Anthony Kenden read his paper. The Soyuz paper was published in the *Journal of the British Interplanetary Society* (*JBIS*) in October 1983 with the Vostok-Voskhod paper following in January 1985.

Tackling manned spaceflight

I was still unsettled about the history of the Soviet manned space programme, and after speaking with David Shayler (publisher of the *Zenit* newsletter) I approached Salamander Books in London with an idea for a book that would not look simply at the cosmonauts but would be a technical assessment of the Soviet manned space programme – analysing the missions, the designs of the spacecraft and the launch vehicles. Salamander liked the idea and *The Soviet Manned Space Programme* (a reasonably descriptive title) was published on 18 November 1988 in Britain and 21 December in America. The Soviets kindly laid on some publicity, as the first (and as events would transpire the only) flight of the Buran space shuttle occurred three days before the book was published and I was able to get some publicity from the resulting TV interviews.

6.4: Presenting a copy of *The Soviet Manned Space Programme* to cosmonaut Yuri Romanenko in 1989.

In the meantime, Anthony Kenden had died due to a car accident in 1987, shortly before he was scheduled to co-chair that June's BIS Technical Forum. The British edition of the book was dedicated to his memory, but sadly the US edition did not carry the dedication.

EAST MEETS WEST

As well as studying the Soviet space programme, I had maintained an interest in the Chinese programme, about which even less information was normally forthcoming. My first major article reviewing the Chinese space programme appeared in *JBIS* for May 1984. By this time, I was also writing about the Soviet photo-reconnaissance satellite programme and was submitting a short-lived series 'The Soviet Space Year' annually to *JBIS*.

I had also gained some experience on radio. While still in Bradford I started to appear as part of the Bradford Astronomical Society's monthly 'spot' on Pennine Radio during 1977. After moving to London, I was contacted by the BBC World Service for a radio interview in April 1984 about the forthcoming flight of an Indian cosmonaut to the Salyut 7 orbital station.

In the latter part of 1987 I was approached by Gerry Webb, who had set up the company Commercial Space Technologies (CST) with Alan Bond of HOTOL and more recently Skylon fame. Gerry had the idea of trying to market the Soviet space programme (which was then just starting to open up commercially) in the West. I joined the company in December 1987 with the idea being that I would be writing technical reports for the company's clients. Working full-time for CST also meant that I would get to travel overseas. My first such trip was with Gerry to the 'Space Commerce 88' conference being held in Montreaux, Switzerland, in February 1988, and it was the first time that I was amongst full aerospace professionals and being treated as an equal. Of course, I was a regular visitor to the Glavkosmos stand – the company set up by the Soviets to market their space programme. It was then that I came face-to-face with Soviet bureaucracy for the first time. No, the Glavkosmos representative would not tell you what literature he had; I had to ask and then he'd see if he had what I wanted. My suggestion that he let me into the filing section was met with a straight face and "*NYET*" which was clearly in capital letters.

A trip to the "Evil Empire"

On returning from the conference, Gerry Webb and I had a meeting with some people from the BBC who were working on a programme for its *Antenna* science series. In the summer of 1988 the Soviets were planning to launch two spacecraft which were to rendezvous with and land small craft on Phobos, the larger of the two small moons of Mars. The *Antenna* team wanted to do a feature about this, and Gerry arranged for me to be added to a BBC crew which was already scheduled to travel to Moscow. On 19 March I left for Moscow on an Aeroflot flight whose arrival was "interesting" by virtue of landing in thick fog and a snowstorm. I got a cab to the Hotel Belgrad Dva, not the best in the city but it was the only one available at such short notice.

I was given instructions that the following morning I should travel by cab to the Intourist Hotel where I would meet the BBC crew for a decent breakfast. Of course, there are all sorts of stories about a person's room being bugged while in Moscow, and all that I can say is that although there were no electrical leads from the table next to the bed, when the table moved there was an electrical crackle!

Accidental spies!

My first full day in Moscow – the Sunday – was actually a day off work for the TV crew who had been filming in Leningrad the previous week. So Tessa Livingston and David Dugan, the prducers of the programme, and I headed off in the snow to find the National Economics Exhibition which housed the Kosmos Pavilion that contained spacecraft mockups, rockets engines, etc. I was introduced to what I was assured was the Russian tradition of walking knee-deep in snow while eating ice cream. Well, from the number of other people doing it, it was clearly done by most people – even if the ice cream manufacturers had a lot to learn from Western products.

We found the Kosmos Pavilion easily enough. On seeing that it was locked at the front, the three of us went around the back and found part of the building which was open to the public. The BBC producer happened on a door that opened into the

6.5: Outside the Cosmos Pavilion of the Soviet Economic Achievements Exhibition in March 1988.

main Pavilion, so with quick looks over our shoulders we slipped through and spent about 45 minutes examining the spacecraft before the guards found us in there. Thankfully, the two words "British Television" made sure that we weren't taken for a trip to the Lubyanka and we soon arranged to return to film it.

Our KGB minder arrives

Monday morning I met our minder Yuri – supposedly from a Moscow radio station but actually KGB. Our first task was to visit Red Square and this is when I realised that my role in the programme was not to be the off-camera consultant but to be the on-camera presenter. This was a minor point not previously made clear to me! It was interesting that within two minutes of the TV camera being set up, sets of policemen from each corner of Red Square arrived to see what we were doing. Yuri flashed his "Moscow Radio" card at them and they left us alone.

In the afternoon we returned to the Kosmos Pavilion, this time as official guests. This was when I was faced with my first instance of been filmed *whilst* talking. We were where the Luna 16, Mars 3 and Venera 9 mockups were displayed, along with some earlier lunar and planetary spacecraft. "OK, Phil, if you can give us the history of Soviet lunar and planetary exploration in three minutes without a break, then that will be fine". We must have had a dozen outtakes. Talk about a baptism of fire.

Tuesday was the Space Research Institute, IKI, where the two Fobos spacecraft had been built. (The Russian spelling of Phobos transliterates to 'Fobos'.) Both had been sent to Baykonur for launch in July, but we were able to speak with some of the people who had designed the experiments carried, and visit the "clean room" which contained an open tube of superglue that would have seriously perturbed anyone at the Jet Propulsion Laboratory in the United States. Inside the building there was ivy growing over the walls. Outside, all of the windows were fitted with infrared sensors. Security was extremely tight at IKI, there was nowhere to get any food but we were able to get some "coffee". The one piece of humour came when I was interviewing the deputy director, Albert Galeyev, whose English was perfect. I had been told to try and ask a humorous question at the end, so I asked if the Soviet interest in Mars was because its nickname was the "Red Planet". Immediately he came back: "You might as well ask the Americans if they study space because the sky is blue."

Drinking 'Vodka' at the IBMP

Wednesday and Thursday were totally different, being at the Institute of Biomedical Problems (IBMP) where the experiments with dogs and monkeys in space had been put together. There were control rooms for monitoring the health of the crew of the Mir space station. I had met the director Oleg Gazenko in 1987 and he remembered. He shocked the Institute's official public relations man by greeting us in person and reminiscing about our discussions over beer in London.

The situation at the Institute was totally different compared with that at IKI: we had a first-class lunch both days, and although we were clearly being watched there

6.6: With cosmonaut Vladimir Solovyov in mission control.

was no feeling of that. There was one humorous moment. Each morning one of the crew would fill up a bottle with filtered water ("this filter can purify the Ganges") to overcome the problems with the tap water. My bottle bore a label for vodka. So, the Russians would see me taking long swigs from this vodka bottle without any sign of being affected, which was a sure way to earn their respect.

Friday was both the highlight and the wind-down. The morning was filming on the Moscow underground, but Yuri announced, "This afternoon we are going to the Mir space station control centre." "Oh," I said, "we're going to Kaliningrad?" "No, we're going to the Mir control centre." That afternoon we were on the Moscow ring road and I pointed out a sign to Yuri: "K-A-L-I-N-I-N-G-R-A-D". "Oh," he said. The van took one turn to the right and one to the left, and we were outside the control centre and I was getting exited.

We met Vladimir Solovyov, a deputy director who had flown two space missions, including the first visit to Mir in 1986. I elected to stay in the upper visitor's section, taking photographs with Solovyov while the camera crew went down to the control room floor. While we were there, the redocking of Soyuz-TM 4 with Mir was being reshown: it had actually happened while I was at Space Commerce 88 and this was the first time I had seen the footage. We interviewed Solovyov in one of the smaller control rooms before being given "mission controller badges" as we left. With that, the main work was over.

That evening Yuri asked how I had known that we were going to Kaliningrad. I explained that the control centre had opened in time for the Apollo-Soyuz mission in 1975 and American visitors had reported it was located at a place called Kaliningrad. I am certain that Yuri knew all along where we were going, but he wasn't allowed to confirm it.

Finally, Saturday was my return to London, having survived my trip to what had been deemed to be the "Evil Empire". Despite rumours, I did *not* emulate the Pope and kiss the ground on my return. The BBC aired the programme on 18 May and it marked the start of a foray into the role of being a "talking head" on television.

Becoming a 'space expert'

After all that excitement, life returned to something approaching normality. In July 1988 I was seconded for some BBC News work with the launches of the two Fobos missions. September saw some excitement when there were recovery problems with Soyuz-TM 5: the two cosmonauts had undocked from Mir and separated from their orbital module in order to prepare for retrofire. However, a combination of a sensor problem and a software issue meant that retrofire did not occur. The media had not realised the seriousness of the situation – without the docking system carried on the orbital module there was no way of redocking with Mir and no rescue craft could be launched. I alerted James Wilkinson at the BBC to this problem (ITN always used Geoff Perry, so we almost became friendly rivals) and when I was called in for an interview I took a Soyuz model with me. After that, I was invited in to the London offices of some US news companies to provide follow-on interviews. One of them specifically asked: "You *will* have that model with you, won't you?" Later someone

6.7: With the Soyuz-TM 9 cosmonauts Alexandr Balandin and Anatoli Solovyov, plus a now-famous plastic model of Soyuz.

called from the United States and asked how much the model cost. My answer was, "About £5, plus the return air fare to Moscow!"

BBC provides a 'scoop'

In 1990 I was working once more with David Dugan of Windfall Films, this time purely as a consultant for a three-part television documentary which was shown by the BBC in November-December that year as *Red Star in Orbit* (in the United States it was *The Russian Right Stuff*). The film crew had unprecedented access (for that time) to the people behind the Soviet manned lunar programme and the cosmonauts who had been scheduled to fly missions around the Moon and down to the surface. Cosmonaut Alexei Leonov, the first man to walk in space in 1965, was particularly eloquent and had a wonderfully humorous and relaxed style in relating anecdotes.

One of the major people interviewed for the programme was Vasily Mishin who was Korolev's principal deputy in the early years and after Korolev's death became head of OKB-1, the Korolev design bureau. Of particular interest was his freehand sketching of the N-1 rocket that had only recently been declassified. Windfall Films also gained access to the museum at the Moscow Aviation Institute, which had just been revealed to contain the remains of spacecraft intended to carry cosmonauts to the Moon – including the lunar lander. With the official "minder" blindsided, David Dugan, his a camera operating, was able to open the hatch of the lunar module and gain the first glimpse inside the spacecraft. When I saw the rushes from the filming my comment was that I wouldn't feel safe in that thing while it was being taken to the launch pad – never mind attempting to fly it down to the lunar surface!

With the agreement of Windfall Films, I was able to put together two articles for *Jane's Soviet Intelligence Review* that featured a lot of then-new information and of course pictures taken by the TV crew: 'Plans, Politics and Personalities' in December 1990 and 'Trying to Fly the Machines' in January 1991. These included my version of Mishin's sketch (he destroyed his after the filming), showing the tapered design of the N-1 and the then-new information that its propellant tanks were spherical. While these articles had new information in the English language, their publication was pre-empted by Mishin's own account 'Why Didn't We Land On The Moon?' in Russian as the December 1990 issue of the journal *Znanye*.

Seeing a shuttle launch for free

After a major reorganisation of Commercial Space Technologies at the beginning of 1991, I opted to start working on a freelance basis. The highlight of the year was not specifically work-related, though.

Since 1986 I had been contributing to the *TRW Space Log* and was surprised to get a call from its editor Tina Thompson in 1989 asking if I fancied a trip to see a shuttle launch. The deal was that if I got myself to the United States, TRW would then pick up the bill as a 'thank-you' for my *Space Log* contributions. The launch was to be of Atlantis on the STS-37 mission to place the Gamma Ray Observatory into orbit. I arrived late on 2 April 1991 and I spent the first two days enjoying the sights at the Kennedy Space Center. Raiding as much of the library as I could, I had pretty-well filled the trunk of Tina's rental car with papers.

6.8: Phil Clark is "made to feel small" by the Saturn V rocket.

On the afternoon of 4 April we had our official trip around KSC as a hired coach drove us past the Vehicle Assembly Building, by the side of the crawler way to Pad 39B where Atlantis was sitting ready for launch the following morning. After a stop at Pad 39B for photos we passed Pad 39A, with Discovery encased in its protective shroud, before stopping at the Saturn V display – one of the few things that can make me look small. The launch on the morning of 5 April was perfect. For me the surprise and disappointment was how soon everything was over. After 90 minutes or so all of the visitors had gone and the TRW staff, along with everyone else, was packing up. I was stood there thinking: "Hey, we've just had a shuttle launch and we've already forgotten it!"

THE END OF THE USSR

The end of 1991 was to have a major impact on the consultancy work I was doing. The Soviet Union ended and, far more rapidly than before, things opened up. Now Westerners could simply talk directly to Russians about doing space business and the need for 'consultants' declined.

The end of 1992 brought a change which was good for my situation. After trying to work on a "commercial" basis the *Tables of Earth Satellites* as published initially by the RAE and subsequently by the Defence Research Agency (DRA), was unable to cover its costs and ceased publication. By this time, of course, I was utilising the Two-Line Orbital Elements for my analyses and the DRA kindly let me have a copy of their distribution list. I decided to try publishing a replacement called *Worldwide Satellite Launches* (WWSL) on a monthly basis, the plan being to kick off with the January 1993 launches. I sent out a draft of the planned format which was similar to that of the *Tables of Earth Satellites* but with more orbital data and supplementary information. I was able to build up a good circulation for WWSL, and that, together with articles and other writing, kept me in business.

Occasionally I would be asked to do studies for specific clients. One was of the Chinese recoverable satellite programme in 1994. And then in 1997 I was asked to prepare a directory of the different types of Russian photo-reconnaissance satellites. The latter contract was a result of someone seeing the database software that I had written which reflected the numerous classifications and the physical details of the satellites that were starting to 'leak out' in Russian literature. The software has been much updated over the years as more information became available, and the number of databases being interrogated for different enquiries has increased.

I continued to write about the Russian and Chinese space programmes, and also appeared on radio and television (usually BBC World television and News 24) as a space commentator. When I met some Chinese representatives at the Farnborough Air Show in 1998 I was told that they were always pleased to see my interviews – they might not like me revealing as much as I did about their programme but they liked that I was always fair and non-political in what I said.

Consultancy work dries up

By the millennium the Russian space programme had become an 'open book', with Western companies forming joint ventures directly with their Russian counterparts. In 1996 the first proper commercial satellite was launched by Russia for a European customer. There was virtually no need for the role of the independent consultant in this climate. At the same time, the Chinese space programme was opening up greatly to the West once more as commercial space launches were undertaken.

As a result of these changes, the need for the services which I had been offering faded away and by late 2001 it was clear that it was no longer possible to make a living from such work. By this time I was living in Hastings, on the south coast of England, and started working within the civil service. I continued the publication of *Worldwide Satellite Launches* and made myself available for radio and occasional television interviews.

In 2001, I started to analyse the Israeli space programme. Very few writers had realised Ofeq 3 was a reconnaissance satellite that was only manoeuvred five years after its launch. I was able to link this analysis to a review of the maiden flight of the EROS satellite (the commercial version of Ofeq). My paper entitled 'Israel's Ofeq 3 and the EROS satellites' appeared in *JBIS* for May-June 2002.

In the last quarter of 2001 I started work on a paper for the British Interplanetary Society – but it was 18 months before I was ready to draw a final line underneath it. The plan was that the paper would bring together all of the numerical analysis I had done over the years about the Soviet lunar programme. A preliminary version of the paper was presented at the Technical Forum of the BIS in May 2002 and I presented the additional analysis at the meeting in June 2003. The paper entitled 'Analysis of Soviet lunar missions' appeared in the May 2004 issue of the *JBIS Space Chronicle* and to-date it is the last major paper that I have prepared.

With the Chinese starting to test fly their Shenzhou spacecraft in November 1999, and further tests being undertaken in January 2001, March-April 2002 and December 2002-January 2003, I prepared a paper which provided a full orbital analysis of these flights, as well as a look-ahead to what might happen. This was published in *JBIS* in May-June 2003. I had previously had two analyses published by *Jane's Intelligence Review*: 'Chinese Designs on the Road to Space' in April 1997 and 'China's Dream Space Countdown is in Sight' (September 1999) – the titles of which were chosen by the *Jane's* editor. Two additional parts of the *JBIS* paper which covered the flight of Shenzhou 5 and then the 'solo' flight of its orbital module were published later.

The last major story for which I provided television commentary and analysis was China's first piloted space mission, Shenzhou 5, in October 2003. By looking at the previous unmanned test flights (the second, third and fourth ones in particular) it was possible to do some 'number crunching' well in advance of the new flight based on the hints that the Chinese were giving, namely it would last approximately a day and both the launch and the landing would occur in local daylight hours. The edition of *Worldwide Satellite Launches* published shortly before the expected launch brought my analysis together. It predicted a flight time of 21 hours 24 minutes with a

6.9: Phil Clark presenting a paper at the BIS Forum.

launch taking place no earlier than 09:00 Beijing Time, 01:00 GMT. This launch time would give a landing at local sunrise, matching the Chinese criteria of launch and landing being in daylight.

Having tipped off the BBC in advance of the launch, I was called into TV Centre to provide coverage of both the launch and the landing for World Television – doing both meant that I would be at the BBC overnight. Launch came at 01:00 GMT on 15 October as predicted, the main orbit circularisation manoeuvre occurred about seven hours after launch as predicted (this prediction was easy because it was how the two preceding unmanned test flights had behaved) and the flight lasted for 21 hours 23 minutes. As always the media considered such predictions to be akin to magic, even though the laws of orbital mechanics are the same whether you are in mission control centres in Houston, Korolev or Beijing – or using a home computer in Hastings.

Writer's block strikes

I continued to writer papers which would be published in *JBIS* until early 2006. I had prepared a second paper reviewing the Israeli *Ofeq* and EROS satellite programmes, and this was presented at a BIS meeting in November 2005. The final version of the paper, 'Israel's Ofeq and EROS satellite programmes', was submitted in early 2006 and finally published in *JBIS/Space Chronicle* in 2008.

After that, something that all writers dread struck me. In mid-2006 I developed a serious writer's block. I had plenty of ideas for papers and started some of them, but was completely unable to finish them. Hence the second Israeli paper is the last one which I have had published to date. I am still hoping – six years later – to somehow get around the writer's block, but only time will tell.

Of course I continued to publish *Worldwide Satellite Launches*, but in September 2009 I contacted Swine Flu and (by coincidence or not) over the space of around two days the sight of my right eye faded away to the point that all I could do was tell if I was in a lighted or dark room. This was later diagnosed as having been caused by my optic nerve ceasing to function and it was a one-way process. At the same time there was a decline in the sight of my left eye to the point that until I got new glasses it was extremely difficult to read anything, whether on paper or a computer screen. It took three months for the medical consultant to conclude that the problem with my left eye was completely separate from that with the right eye. These medical problems meant that I had to cease publication of *Worldwide Satellite Launches* after a run of nearly 17 years.

At the end of May 2010 I had a major shock when Rex Hall died in hospital. We had been close friends for 30 years. Despite my writer's block, Rex's death has made me resolve to produce a book that I have long wanted to write. I could pull together the TLE orbital data which I have, all of the Soviet photo-reconnaissance satellite telemetry analysis of the Kettering Group, and the newly released information from Russia and link this to my own analyses to produce the first book dedicated to the Soviet-Russian photo-reconnaissance satellite programme. The book has the working title *Red Eyes in Orbit* but work is currently (May 2012) very slow.

CONCLUSIONS

Even as I was studying the space programme of an 'enemy' during the Cold War – a fact that caused some people to question my politics – for me it was never anything political. The Soviet space programme was a mystery and I was simply attempting to solve at least part of that mystery. My interest and desire for more analytical tools led me to study mathematics beyond the normal school level all the way to eventually getting my degree.

I was one of a group of analysts of the Soviet space programme and like everyone I had my successes as well as my failures. But one has to expect this – one cannot be right all of the time, as my 1981 Progress prediction spectacularly proved. When we had to work everything out for ourselves, as opposed to nowadays simply looking it up online, it was great fun, there was camaraderie, and many long friendships were forged. Did I personally make a specific difference? Maybe I did, simply because of the breadth of the analysis which I was doing and in the 1970s-1990s the volume of material I was publishing, mainly by the British Interplanetary Society. But it would be wrong for me to make any exaggerated claims. It was fun to go to a TV studio and speak about the launch of a mission to the Mir space station, giving a second-by-second description of what was happening aboard the launch vehicle. But that was only possible because I had been studying the programme for more than 20 years and had retained what I had learned.

With the Russian and Chinese space programmes now being so open, the need for 'space sleuths' on the scale that we had from the late 1960s to the late 1990s has gone away. Of course, there are still programmes like those of Israel, Iran and North Korea where secrecy prevails, but these are 'small fry' compared with working out what the Soviet Union and China were doing in space prior to *glasnost*.

Of course, we are now in a totally different world in terms of spaceflight analysis compared with the early 1970s when I started on this road. Instead of working away in isolation or via slow correspondence, as most of us were doing in the 1970s, we now have everything almost immediately available with real-time discussion groups providing instant analysis on the Internet. While there are still historical mysteries to solve, these are fewer in number with every passing year. If I can overcome writer's block, my first priority is to write *Red Eyes in Orbit* and see if someone will publish it. Then there are still some analytical articles that I would like to prepare – and see if there is a market for them. Only time will tell.

7

Adventures in Star City

by Bert Vis

"How did you ever become a member of that small group of space sleuths," is a question I'm asked every now and then by others who are interested in the history of spaceflight. The truth is that I don't really know, nor do I know what moment in time marked my "joining". Somehow, I feel that this goes for all of us who apparently are members of this small group. I've always seen 'space sleuth' as an honorary title that one can't apply for.

PIECES OF A PUZZLE

At some point you come across your name in a post on the Internet or in a magazine article as part of a short list of names of space historians that are considered people who have done impressive research and/or have discovered unknown facts about the former Soviet space programme. Whether that's justified is for others to decide. To get to that point, some luck is needed, as is diligence, and the will to spend time on the subject. Often, the time will be wasted, but every once in a while, the result will be a small gem that brings just one more piece of the puzzle. Answers to questions usually result in new questions, and getting the answers to those will require further time and effort and raise even more questions.

After the era of *glasnost* and *perestroika* began in the late 1980s, many secrets of the Soviet space programme were revealed in magazines, books and television documentaries. In the early 1990s the late Rex Hall – perhaps the greatest space sleuth of us all – almost sadly told me that he believed all the puzzling was over. That was probably one of the biggest mistakes he made in all his years of researching the Soviet manned space programme in general, and the cosmonaut group in particular. The puzzling was far from over. In fact, in a way, the puzzling was just starting, albeit perhaps in a different way.

While the names of those who were selected in the cosmonaut detachment were slowly revealed in letters from the so-called Information Group of the Yuri Gagarin

7.1: Rex Hall and Bert Vis in Star City with South Korean cosmonaut Soyeon Yi in 2007.

Cosmonaut Training Centre (GCTC) near Moscow, and from Russian magazines such as *Aviatsiya I Kosmonavtika* (*Aeronautics and Cosmonautics*), it was obvious that there was still a lot that had not been told. Not necessarily because it was still considered secret but simply because you hadn't asked for it.

Unknown 'unflowns'

My own specialty, like Rex Hall's, was the cosmonaut detachment. Who were these guys? In what way were they similar or different from those much publicised guys in the US, who had the term hero bestowed upon them even before they began training. At the time, the names of Russian cosmonauts were only revealed after they'd been launched into space. The names of the unflown guys were closely guarded secrets, as indeed was how many of them there were.

In July 1971 a plaque with the names of deceased astronauts and cosmonauts was placed on the lunar surface by Dave Scott and Jim Irwin of Apollo 15. It bore the names of six Soviet cosmonauts: Vladimir Komarov, Yuri Gagarin, Pavel Belyayev, Georgiy Dobrovolsky, Vladislav Volkov and Viktor Patsayev. At the time, due to the secrecy surrounding the cosmonaut detachment, it was not realised that the names Valentin Bondarenko and Grigori Nelyubov ought also to have been on the plaque. Their names would not be revealed until 15 years later. By that time

Nelyubov was already known about, because he appeared on one of the famous Sochi photographs that had been so meticulously studied by Jim Oberg. Jim had managed to find several versions of the photograph. On one, the man who later turned out to be Nelyubov was evident, but in other versions he had been retouched out of the picture and replaced by either a bush or a staircase. It was fairly obvious that this man was a cosmonaut who had not flown, and thus had to be erased. Finding out what had happened to him was frustrated until articles in *Izvestia* in April 1986 not only gave the names of the eight missing men from Gagarin's selection group, but also their fates.

Moscow's new 'Information Group'

From then on, the names of other cosmonauts also slowly but surely became known. GCTC's Information Group was bombarded with letters from the small group of Western researchers asking all sorts of detailed questions regarding the cosmonauts. Some were answered, some were not, although it never became clear what would be the trigger for Mr. Yegupov, the "Head of the Information Group" to write back. In later years, it became clear that at that time Yegupov wasn't just the head, he *was* the Information Group!

If one thing became clear in those days, it was that you needed to be precise in the way you asked your question: questions were answered, but no extra information was given. At one point, I asked for the names of Valentina Tereshkova's backups, and was told they were Valentina Ponomaryova and Irina Solovyova. The names of the other two members of Tereshkova's selection group were not given [1]. At that time, the biggest success was the full list of backup crews from Gagarin up to Soyuz-TM 5, revealing six new cosmonaut names, among them those of what was said to have been the Soyuz 13 backup crew of Lev Vorobyov and Valeriy Yazdovskiy [2]. Later, it would become known that these two men had been the intended prime crew and had been replaced by the real backups only days before the launch.

The list was published in *Spaceflight* magazine, which was at that time pretty much the central point from where all of the new information was shared with those interested in manned spaceflight [3]. The same issue carried two letters from one of the best-known Russian members of the unofficial group of space sleuths, Vadim Molchanov. One of his letters revealed the names of the cosmonauts of the 1963 selection group. Later, Dutch sleuth Anne van den Berg would manage to get the full list of the cosmonauts selected in 1965, which included many new names because only six of the 22 selected that year actually flew in space.

Rumoured cosmonaut names

It was remarkable to discover that many of the rumoured names were not among the ones released. Over the years, various sources had published sensationalist stories of cosmonauts who died making spaceflights which went terribly wrong. What was the story here? The best way to find out was to simply ask, so in April 1988 a new letter went to Yegupov asking him for information on all those people who had reportedly been cosmonauts in the 1960s and 1970s – Vetrov, Voronov, Vinogradov, Asanin, Obraztsov, Vavkin, Kornyev, Bogdashevsky, Raushenbakh, Zhukov, Barsukov, Bondarev, Antoshenko and Ilyin.

It took six months for Yegupov to reply, and he denied any of these men had ever been cosmonauts with the exception of Anatoliy Fyodorovich Voronov, who was a member of the second group of cosmonauts that was selected in 1963 [4]. The name had been published by the French news agency AFP in 1972 as being in training as a future Salyut commander together with that of Vinogradov, who was said not to have been a cosmonaut. But Voronov was a common name in Russia and the possibility of a chance hit was generally accepted.

A surprising photograph arrives

In those days, autograph hunters would write to the GCTC to get an autograph or other reply from the flown cosmonauts. Trying to find biographical information, in particular information on what activities he had carried out during his cosmonaut

7.2: Unflown cosmonaut Anatoliy Voronov.

career, I decided to simply write to Voronov care of the GCTC. It was a long shot, but it only cost me a stamp and some time to come up with a letter asking him for a portrait photo and answers to some carefully selected questions. The worst thing that might happen was not getting a reply. But that was something I was already used to, since a certain percentage of letters to cosmonauts or the GCTC Information Group went unanswered. I had all but forgotten sending the letter when in May 1989 I received one of those by-then well-known envelopes from Russia. It contained an autographed official portrait photo of none other than Anatoliy Voronov. For the very first time we had an official portrait of an unflown Soviet cosmonaut, and I displayed it in June at the British Interplanetary Society's Forum on the Soviet space programme.

Of course, now that the contact had been made, the next task was to build on it. Although the portrait was very welcome, Voronov had not actually answered any of my questions, so a follow-up letter was written to thank him for the photo and to ask whether he would be willing to send me some biographical information, in particular on what he'd done after his selection into the cosmonaut detachment. His name had not been included in the list of backup crews I'd received earlier from Yegupov. The letter went out in July and the reply came three months later. Besides giving basic biographical data, all of which was of course new, Voronov explained that between 1967 and 1970 he'd been a member of the cosmonaut group that had prepared for a lunar mission. From 1970 he'd trained with Aleksey Gubarev and Vitaly Sevastyanov for a Salyut mission, but when in the aftermath of the Soyuz 11 accident the crew size was reduced to two, he had lost his seat. After that he had trained for some five years with Vladimir Lyakhov on Soyuz-T, but in the end had been removed for medical reasons. At the time of writing, he was working for the state centre 'Priroda' that was conducting research on Earth's resources using space capabilities [5].

As said, answers always raise new questions, so shortly after receiving this letter a new one went out asking for more details on his involvement in the lunar programme and for the names of others that had been involved. Once again, a reply was received in which a list of 25 names was given of cosmonauts who had been members of the lunar group [6].

Pre-Internet networks

In that pre-Internet era, all of this information was quickly Xeroxed and shared with others in our small group of space sleuths. And by combining all these snippets we gradually managed to obtain a better view of the big picture. It was also interesting to realise what information the Russians weren't revealing to us. Apparently, there were still subjects they would not talk about, and of course these were of particular interest to us. I myself concentrated on the cosmonauts and their careers, and whenever new names were released I'd write a letter hoping to get a response. And in some cases I did. I began to concentrate on the unflown cosmonauts, not only because so little was known about them but also because I thought they deserved more attention than they were getting from the Russians themselves. Besides, once someone flew, the official press would publish a portrait as well as biographical information, and even though it was apparent that there was a lot more to tell about these cosmonauts, the fact that hardly anything at all was known about these 'unflowns' made them that much more interesting to me.

MEETING THE COSMONAUTS

In July 1990 a first big interview opportunity presented itself. The annual congress of the Association of Space Explorers was to be held in Groningen, in the Netherlands. It was thought that over twenty Russian cosmonauts would attend and, together with Gordon Hooper, who was also trying to find out information on cosmonauts for his monumental (especially at that time) *The Soviet Cosmonaut Team* [7], I went to see whether I could get to talk with some of them. It was a long shot, as we had no idea how easily these cosmonauts could be approached, and how willing they might be to answer questions.

Revealing interviews

It was a revelation. We managed to interview Sergey Krikalev, Oleg Makarov, Igor Volk, Valeri Polyakov, Svetlana Savitskaya and last but not least Vitaly Sevastyanov.

Gordon and I had put some effort into how we should ask certain questions. For example, it had long been assumed that Soyuz 7 and Soyuz 8 were to have docked in space but this had not happened. We decided to try and bluff our way to an answer, and told Sevastyanov that we knew the docking attempt had failed but didn't fully understand what had gone wrong. He fell for the trick and started to explain that a fault in the range-finding equipment had prevented the docking. That was great! Not only did we now have a more or less official confirmation that a docking had indeed been intended, we also knew what had gone wrong. It was our first scoop.

Another one was the first portrait that I was able to publish of Yelena Kondakova, a newly selected cosmonaut reported to be undergoing her basic cosmonaut training with three other candidates (Budarin, Poleshchuk and Usachev). She would have had no business at the ASE if she hadn't been married to cosmonaut Valeri Ryumin and was simply accompanying her husband. She hadn't counted on being recognised by us and photographed.

By the end of the week, we had quite a lot of new data that was incorporated into Gordon's book and was published in three articles in *Spaceflight News* [8]. I had by then concluded that this was such a great opportunity to interview cosmonauts that I would attend the next ASE congress in Berlin in October 1991. I went alone, and was able to get extensive interviews with cosmonauts Afanasyev, Sevastyanov, Yegorov, Serebrov, Balandin, Strekalov, Kovalyonok, Manarov, Zudov, Dyomin, Gorbatko and Volk. Two of them, Sevastyanov and Volk, were encores. With the results of the previous year in mind, I had prepared much better and once again both men gave lots of interesting new facts. In the case of Sevastyanov, he spoke extensively about

7.3: Bert Vis interviews Oleg Makarov during the Association of Space Explorers congress in 1993.

the crewing history for Salyut 1, and of the cosmonaut group that had trained for lunar missions.

Travels across Europe

Based on my experiences at the 1990 ASE congress, I had also decided to take my chances and attempt to secure individual interviews, however brief, with cosmonauts that I knew would be travelling to places that I could relatively easily reach. And so I interviewed Oleg Atkov in Noordwijk, the Netherlands, when he was visiting ESA's ESTEC. Our talk lasted only about ten minutes but I gained a better understanding of how the Institute of Medical and Biological Problems (IMBP) cosmonaut group was formed and even got some new names of finalists for the selection groups.

Having heard that Valeri Ryumin was to attend the Air and Space Salon at Le Bourget, I drove 500 kilometres to Paris. It was a long shot, but even if he refused to be interviewed I could still visit the various pavilions. In fact, we talked for half an hour – twice as long as we had initially agreed upon. All this encouraged me to keep attending the ASE congresses. Some cosmonauts would recognise me from earlier meetings and sit with me again to answer my never-ending questions. I followed the congress to Washington (1992), Vienna (1993), Warsaw (1995), Canada (1996), Brussels (1998), Bucharest (1999) and Madrid (2000). By then, there was so much information on the Soviet-Russian manned space programme coming out through other sources that I was no longer primarily trying to get names of cosmonauts or crewing information. Instead, I was concentrating on other facets of the programme, such as training.

LETTERS FROM MOSCOW

By the early 1990s I was still writing to every new name that was revealed. Some of them replied, and whilst some only sent a portrait, this was still welcome because it was another step toward completing the overview of the cosmonaut detachment. A few cosmonauts took the trouble to write letters, and even the brief letters contained interesting information. Lev Vorobyov (who had been identified in Yegupov's letter as the Soyuz 13 backup commander) wrote that he did indeed train for that mission but had not been the backup. He apologised for not going into detail, saying that he "didn't like the subject". Many years later, when Vorobyov was no longer merely a name and a portrait but had become a close friend, he would tell me his story.

A personal friendship is born

One of the cosmonauts who replied with a long letter was Sergey Gaydukov. Selected in 1967, he was assigned to flights of the military Almaz space station and trained as a member of a large group of cosmonauts for that specific programme. When it was

7.4: Unflown cosmonaut Sergey Gaydukov.

cancelled he was not transferred to Soyuz-Salyut but served as a communicator on one of the tracking ships.

In 1988 he suffered a cerebral haemorrhage that paralysed the left side of his body. No longer able to work, it turned out that my letters were a welcome distraction from his everyday life and he very patiently wrote long letters to me in reply to my letters with ever more questions about his days in the cosmonaut team. His handwriting had obviously suffered from his condition, but I located a fellow space sleuth who was fluent in Russian and he learned to read Gaydukov's handwriting. That person was Chris van den Berg, a radio amateur who for years had been listening in to the air-to-ground radio traffic between cosmonauts and mission control near Moscow. Over the following years Chris would decipher and translate Gaydukov's letters and assist me in corresponding with other cosmonauts – none of whom spoke English at that time.

Invitation to Star City!

It was after a couple of letters that Gaydukov did the (at that time) unthinkable – he invited me to come to Star City to meet with him and his friend, fellow cosmonaut Gennadiy Kolesnikov. I replied that if this was a serious offer I would certainly come to Russia, and soon I had confirmation that once I was in Moscow he would send his daughter Irina and Kolesnikov to pick me up.

It sounded almost too good to be true, but in October 1991 I travelled to Moscow with a regular tourist group, solving any visa problems that might have arisen. Once I let Gaydukov know in which hotel I was staying, we agreed that I'd be picked up the following morning. At 9 a.m. I was waiting outside the hotel when a man came up to me asking if I was Bert Vis. I was puzzled, because this guy wasn't Kolesnikov. He introduced himself as Vadim Molchanov, my fellow space sleuth from Tula. Of course! I had written to him about my coming to Moscow and we'd agreed to meet, but that would be later in the week. Now he was here instead of Irina Gaydukova to pick me up. Another surprise came when I realised that Vadim hadn't come with Kolesnikov, who was said to be busy organising the final details of my visit to the training centre. When we got into the big Volga automobile, I was introduced to none other than Lev Vorobyov, the man who had been kicked off the Soyuz 13 flight back in 1973.

Upon arriving at the Star City gate, we were met by Irina Gaydukova, Gennadiy Kolesnikov and a guide from the GCTC museum, Yelena Yesina, who would act as interpreter. Later, it would turn out that Irina's English was excellent and there was no real need for Yelena's help, but in retrospect I'm happy that she was there because we became very good friends.

Free to photograph anything

Before Kolesnikov led us through the gate into the training centre, I was told that I was free to photograph anything. This was way more than I could ever have hoped for. The first building I was shown was the one that houses the Mir modules, where a

7.5: Walking towards the centrifuge building in 1991. (L-R) Irina Gaydukova, Yelena Yesina, Vadim Molchanov and cosmonaut Gennadiy Kolesnikov.

detailed explanation was given by Kolesnikov. In between questions and answers, I was happily snapping away with my camera, although in that pre-digital era I had to make sure I would still have some film left for the later part of the tour. In those days, it was difficult if not almost impossible to get film in Russia. Running out of it would have meant no more photographs.

After the Mir hall, I was taken to the planetarium where cosmonauts were trained to navigate by the stars. Kolesnikov explained that before the planetarium was built, cosmonauts would be flown to places as far away as Murmansk or Ghana in Africa to study the skies.

Leaving the planetarium, we went to an office in the same building where I was introduced to Vladimir Dzhanibekov, at that time the most experienced cosmonaut with five flights. In those days, meeting a cosmonaut was far less common than it is nowadays, as they weren't travelling to Houston or Western Europe all the time, so I was thrilled when Dzhanibekov told me it was his pleasure to meet me at the request of Sergey Gaydukov. Perhaps I has some questions? Of course I had, and in the next fifteen minutes I asked about the crews he had served on. Until then, few details had been released on the crewing of Soviet spacecraft and I wanted to clarify some points. Crew changes were more common in the Soviet programme than for NASA. Getting the cosmonauts' own stories on the subject would be interesting, I thought.

Because the personnel of the centrifuge facility were on a lunch break, I was first shown around the GCTC museum in the town itself, after which we returned to the training area where I was shown around the centrifuge and the hydrolaboratory used for spacewalk training. I was thrilled. At that time, visiting the GCTC was restricted to VIPs. Not long after that, groups would occasionally be permitted to visit, but only by paying a pretty high fee.

Once the tour was over, I was taken to Gaydukov's home in the second of two large apartment blocks behind the Gagarin statue that was well known from official photographs of cosmonauts laying flowers there. I finally met with my 'pen pal', and thanked him for the unforgettable day which he and Kolesnikov had arranged for me. But it was not over! The Gaydukovs had organised a big dinner and it was as if the prodigal son had returned home. During dinner, it was said that it would be a waste of time to go back to Moscow, and I was invited to stay over for the night. I would have to get official permission for that though, and for that Kolesnikov contacted General Dyatlov, the chief of staff of the GCTC.

Drinking with the cosmonauts

In the meantime, I was expected to visit several other unflown cosmonauts, so I was taken to Pyotr Kolodin's and Lev Vorobyov's places where I got further information on crewing and cosmonaut training. On our way back to the Gaydukovs we ran into yet another of the unflown cosmonauts, Anatoli Dedkov, who, on hearing that I was the guy who had been sending him letters, insisted that I accompany him to his place for a drink. This yielded new crewing information too. My one-day visit was yielding more information than three years of corresponding with the cosmonauts.

When it was time to go to sleep, I was told that I'd be taken back to Moscow at 8 a.m., but I had to be downstairs five minutes early. Aleksey Boroday had heard that I was in town and insisted that he wanted to meet me. As he was having exams that day it would have to be short and would in all probability be just a shaking of hands, but he'd indicated he really wanted to meet and so we did.

When saying goodbye to the Gaydukovs, I thanked them for what had been a visit that I could only have dreamed of all those years. All in all, I had met almost a dozen cosmonauts in less than 24 hour. Of course, I had also met that number at the ASE congresses, but here only two of them had actually flown in space. The others were all unflown, and that was unprecedented because they never visited the West. Only a few years earlier, all their names had actually been state secrets!

The biggest surprise, however, came as I was saying goodbye to my hosts. Sergey Gaydukov told me I should come back the following year and not for a day but for a week. He would arrange for me to stay at his place so I could interview cosmonauts and have a more relaxed look around town. I was thrilled. My second trip to Star City and the GCTC would be in July 1992.

Appearing in 'Who's Who'

Much of the information I managed to collect in those years ended up in Michael Cassutt's *WHO'S WHO IN SPACE* [9]. Copies found their way to Russia, and the unflown cosmonauts were said to be very grateful for the fact that their work was being recognised at last. And indeed by spaceflight enthusiasts in the West too!

When Sergey Gaydukov first invited me to visit him, I had expressed the wish to visit the local cemetery to find the graves of several cosmonauts who had died in the years prior to their names being released.

In June 1990 the Russian magazine *Aviatsiya I Kosmonavtika* published a list of all the military cosmonauts selected up to that point, but there was still information missing or clearly incorrect. One of the things I was looking for were dates of birth and death, and where better to find those than their headstones? However, I had been told that visiting the cemetery would not be possible, although at the time no reason was given.

A SECOND VISIT

It was during that second visit that I again said I'd like to visit the cemetery, and this time I was given the opportunity. I went together with Vadim Molchanov who, like the previous year, had come from his hometown Tula to meet me. As we walked up there, it became apparent why it had not been possible the previous time. Not only was it quite a way, the road was poor and in October the rain and mud would have made the trip difficult. The cemetery was a part of the forest outside the town and one could readily see that it too would be very muddy after several rainy days. July was obviously a much better time. All in all, the walk to and from the cemetery and our search took several hours, and apart from the road conditions during my earlier visit, I simply wouldn't have had enough time.

This second visit provided a treasure trove of new information. By the end of the week, I'd had extensive interviews with thirteen unflown cosmonauts. Four of them would later fly missions to Mir and the International Space Station: Vasili Tsibliyev, Yuri Malenchenko, Talgat Musabayev and Valeri Korzun. The others were from the generation which would now never be assigned to a mission: Aleksandr Petrushenko, Nikolay Porvatkin, Mikhail Burdayev, Nikolay Grekov, Valeriy Illarionov, Valeri Beloborodov, Nikolay Fefelov and Gennadiy Kolesnikov.

Gaydukov and his wife Valeriya arranged for all the older generation cosmonauts to talk with me while Yelena Yesina helped me to get to talk with the younger guys. It was a busy week, with even busier times to follow when I transcribed the interview tapes, distilled the information and incorporated it into crewing lists. It was great fun though.

More personal details revealed

In my letters leading up to this second visit, I had told Gaydukov how thrilled I had been to be able to meet with Dzhanibekov during my first trip. Now that we had more time to sit and talk, Gaydukov explained his special bond with Dzhanibekov. In 1974, Gaydukov's three-year-old son Dmitriy suffered a bad accident and urgently needed a blood transfusion. It turned out that Dzhanibekov was a perfect

7.6: A rare image of the Hydrolaboratory under construction in 1979.

donor. At that time he was in training for the ASTP programme and, although it was pointed out this might disrupt his training, he immediately agreed to be the blood donor. In fact, he donated blood three times. Thanks to that, and to a hard-to-come-by medicine that Valentina Tereshkova managed to obtain, Dmitriy made a full recovery. He went on to become a cosmonaut trainer at the GCTC.

Valeriya Gaydukova organised my interviews very efficiently, and at one point, while I was interviewing one cosmonaut, the next one would already be awaiting his turn! It was especially rewarding to talk with them and save their voices on tape. The first one I interviewed was Petrushenko, and once we'd finished he told me we would speak again on my next visit to Star City. Little did we know that he would pass away only four months later, aged only fifty.

Watching a training session

My trip didn't only consist of interviewing people. Once again I was given the chance to visit the training centre itself, and I witnessed a training session in which Aleksey Boroday practiced manual Soyuz dockings. Afterwards I was taken to the building where the cosmonauts had their offices, and in a conference room I met most of the younger cosmonauts; the people who would become Mir and ISS commanders in the following two decades. Their commander, Anatoliy Berezovoy, was also present, as was one of the older generation of cosmonauts who had seemingly disappeared from the face of the Earth – Valeriy Illarionov. In fact, lacking a good portrait, I failed to recognise him until he told me that he spoke a little English as a result of spending time in the US working as a Capcom during ASTP. It was pretty funny: while I was embarrassed that I hadn't recognised him any sooner, he was thrilled that I knew who he was. I arranged for an interview later in the week. When we eventually sat down, he apologised for the state of his English but said that he hadn't had to use it since ASTP except when he met with Jim Oberg in Star City. In the sauna to be more precise!

It was also during this second visit that I was given a private tour of the mission control centre (TsUP) in Kaliningrad, now renamed to Korolev. The large hall that I entered had a vast mosaic that featured the three demigods of Russian cosmonautics: Tsiolkovsky, Korolev and Gagarin. There, I was welcomed by Vsevolod Latyshev, whose business card said he was "Head of Information Departament" (sic). He gave me the grand tour and showed me the two large control rooms, one for Mir and Soyuz and the other for Buran. The latter was awaiting a resumption of Buran flights. Of course that never happened. These days the room is used to control Soyuz dockings at the International Space Station.

Rare chance to watch a cosmonaut funeral

Nowadays, the number of Westerners in TsUP is pretty large, and NASA and ESA have their own offices there. But in 1992 things were different, and in many ways it was still much more the Soviet Union than present-day Russia. As I had been doing in Star City and the GCTC, I'd been using up roll after roll of film shooting anything that they allowed me to photograph. At the end of the tour, Latyshev asked if I was using any 400 ASA film and if so, could he by any chance have one since he couldn't obtain it in Russia. As I didn't use 400 ASA film, I offered him a 200 ASA which he accepted. In return, he went to his office and got me an autographed portrait of Helen Sharman.

In Star City, the Gaydukovs kept inviting unflown cosmonauts to come to their place to meet with me. Many did, but not all. Aleksandr Kramarenko flatly refused, saying that talking with a Westerner would not be good for his career. Gaydukov and Kolesnikov informed me that he was paranoid, but some day he would sit with me so that I could put my questions. He never did though, and on 13 April 2002 he passed away during one of my stays in Star City. I was present when he was lying in state in the Dom Kosmonavtov, the town's central building that houses the museum, library and theatre, and where expositions and official events are held. It was impressive to see the honour guard made up of active and former cosmonauts, and young soldiers with machine guns. The next day I had the rare opportunity to attend a cosmonaut's funeral in the cemetery of Star City.

However, most cosmonauts who were asked consented to speak with me, and the Gaydukovs, especially Valeriya, made sure that my time was used to the utmost. At one point, she persuaded Nikolay Fefelov to come to her house for an interview even

7.7: Cosmonaut Aleksandr Kramarenko's honour guard.

though it was approaching midnight! But Fefelov granted all the time we needed. An amateur artist, he was involved in several of the unofficial wall-newspapers that were made before missions, *NEPTUN*, *KOSMONAVT* and *APOGEE*. A single copy would be produced and displayed on a bulletin board. They were said to be full of unknown pictures, jokes and nice cartoons. I indicated that I'd love to see them, but to this day I haven't. Even an official request to be able to see and photograph them was refused by the director of the museum. I don't know the reason. Perhaps the jokes on them are considered inappropriate for public disclosure!

When the time came to leave, it was obvious that I was expected to return the next year, and it soon became clear that I could visit Star City on an annual basis, or even more often if I desired.

MAKING AN IMPRESSION

During my 1993 visit I once again interviewed ten cosmonauts, of whom seven were unflown. I was told by Lev Vorobyov that during my previous visit I had made quite an impression on people when, on seeing two flown cosmonauts together with Dmitri Zaikin, I had said that I really wanted to talk with Zaikin. In fact, I was told several other unflown cosmonauts who had heard about this had told Gaydukov that when I paid my next visit they would like to talk to me about their careers. At that time, the Russian media took little interest in the people who had been selected as cosmonauts but for whatever reason had never been able to fly in space.

This way, I managed to meet and talk with lots of unflown cosmonauts. For some of them my interview was probably the only one they ever gave, as they passed away not long after, before the Russians themselves, and in particular the people from the Russian space magazine *Novosti Kosmonavtiki* started to show any interest.

That year also included another visit to the training centre itself, and as previously, I photographed everything I thought was of interest, whether I had done it before or not. This way, I could compare the photos and detect changes to buildings, simulators and other hardware. I still do this, and it has proven useful in dating the pictures that appear in books and magazines and on the Internet.

It was during this visit that I was shown around the medical department, and was offered a ride on the infamous rotating chair. I accepted, and a few minutes of sitting in the chair and making the head movements required of cosmonauts was enough to cause me to be nauseous for over an hour. The doctors had a ball though, and they proudly showed me their guest-book and invited me to sign it. Of course, I inspected it and found a few well-known autographs of European astronauts like Klaus-Dietrich Flade's, who wrote about the doctor who had tested him, dubbing him "Sadist #1", with "Sadist #2" being his sports instructor. Tim Mace, who was Helen Sharman's backup, wrote that he was looking forward to the day that he could strap the doctor in the chair and be at the controls himself!

Interviewing a reluctant backup

One of my interviews was with Lev Vorobyov and I asked him about his replacement as prime commander of the Soyuz 13 mission. Referring to his logbook, he said the original crews were formed on 2 August 1973. Shortly before the mission the prime and backup crews left Star City for Baykonur with Vorobyov and his flight-engineer, Valeriy Yazdovskiy, still as the prime crew. In Baykonur it was decided to switch the crews. When I asked who actually made that decision, Vorobyov said that it was the chairman of the State Commission, Kerim Kerimov, Georgiy Beregovoy and chief designer Vasily Mishin. It hadn't even been a decision of the State Commission, as Beregovoy wasn't a member of that. But Beregovoy had never liked Vorobyov and, as Vorobyov explained, a lot of people were against Yazdovskiy, who was described as a strong-willed person. The same went for Vorobyov, however. The previous year, he had told me that when they arrived in Star City to begin training they had met the twelve active but still unflown members of Gagarin's group – and decided that the promise that they'd soon be assigned a mission had been an empty one.

The members of the second selection group were actually older than those of the first, and were graduates of military academies. As such, they'd had excellent career perspectives in their former roles. The new cosmonaut candidates began to grumble. This was reported to the air force headquarters by the staff of the cosmonaut training centre. In November of 1963 the air force's deputy commander-in-chief, Marshall of Aviation Rudenko summoned the members of the group and explained that they must be patient. Vorobyov then asked what Rudenko thought they should do at the training centre. Rudenko suggested they read *The Andromeda Nebula*, which was a popular science-fiction novel at the time. Perhaps it had been said in jest, but Vorobyov told Rudenko flatly that they weren't little boys but experienced pilots and were not even flying planes at that time.

There had just been a transition of command at the training centre, with Major-General Nikolay Kuznetsov becoming the commander. Kuznetsov would later tell Vorobyov what happened after Vorobyov had given Rudenko a piece of his mind. When the cosmonauts left Rudenko's office, Rudenko had told Kuznetsov that the wayward Vorobyov couldn't stay in the cosmonaut group and that Kuznetsov must dismiss him. Two things saved Vorobyov's cosmonaut career: the fact that he was a Military Pilot First Class, but more importantly the fact that Kuznetsov had no wish to dismiss him. He had taken over command from Lieutenant-General Odintsov, a very unpopular commander who was replaced after ten months. Only days in office, Kuznetsov decided he would not start his tenure as commander with the dismissal of a cosmonaut. Over the years, Vorobyov and Kuznetsov would become close friends.

But in December 1973 Kuznetsov was no longer the commander. Beregovoy had been appointed in June 1972. As noted, Beregovoy didn't like Vorobyov while many others didn't like Yazdovskiy. Three times, Vorobyov was asked to write a negative report on Yazdovskiy and ask for him to be replaced, but Vorobyov refused. Now, days before the launch of Soyuz 13, the three officials came together and decided to exchange the prime and backup crews. Their reasoning will, in all probability, never be cleared up because all three have passed away. Whatever that reason, Vorobyov would never forgive them.

The official reason for grounding the prime crew was psychological differences between Vorobyov and Yazdovskiy. However, both would assure me there was no animosity between them whatsoever. In 2000, I was able to talk with Yazdovskiy. He and Vorobyov greeted one another like old friends [10]. In our interview, Yazdovskiy corroborated Vorobyov's account of the events in December 1973.

Third time lucky

By sheer good luck, my third visit to Star City enabled me to attend the homecoming ceremony of the thirteenth Mir expedition crew of Gennadiy Manakov and Aleksandr Poleshchuk, along with Jean-Pierre Haigneré who had made a 21-day visit during the handover.

During my trips I kept collecting information about how the GCTC was organised. With this, I managed to make a rough organisation chart of the training centre, giving a pretty complete picture of the situation in mid-1993, but obviously the organisation was dynamic and within a few months my chart was out of date.

"That's secret!"

With the many pictures and notes that I made, and using a rough sketch which I had managed to get, I was also able to make a map of Star City and the training centre. In 2005 my map was published in *Russia's Cosmonauts* [11]. When I later presented the then-commander of the training centre, Vasili Tsibliyev, with a copy of the book as a 'thank-you' for everything the staff of the centre had done for me over the years, he looked at the map and said, "That's secret!" I smiled and forgot about that remark until I presented another copy to his second in command, Valeri Korzun,

7.8: An almost forgotten model of an abandoned 1980s' renovation of Star City.

several days later. He too looked at the map and told me that it was secret information. Apparently both weren't joking. With dozens of NASA personnel walking around in Star City and the GCTC, and with more public sources becoming available (Google Earth in particular), I couldn't imagine how a simple map could still be deemed sensitive or even secret.

But I'd had enough experiences not to be too surprised. I vividly remembered a visit to NPO Mashinostroyeniya, the former OKB-52 that was managed by Chief Designer Vladimir Chelomey. OKB-52 had been Korolev's principal competitor, and had designed the Almaz military space station, two of which were flown as Salyut 3 and Salyut 5. When I visited the place, I wasn't allowed to take any photographs. It was all said to be a military secret. Much to my surprise, however, the very hall that I had visited, as well as the Almaz station and transport ship that it housed, were shown in excruciating detail the very next evening in a television documentary on the history of the design bureau.

There were also cosmonauts who would keep stuff secret from me. For example, Gennadi Sarafanov, who had been launched to occupy Salyut 3 and then been obliged to return after he was unable to dock with the station, told me flatly that he would not discuss the project. If there was something I desired to know, then I could consult one of the publications which had revealed details. Pavel Popovich, who had at one time been involved in a military variant of the Soyuz called 7K-VI crossed his arms in front of him as soon as I mentioned the name, saying he would not speak about it. Even my attempt to get him to talk just about the crewing for the project didn't make him change his mind.

Fellow travellers

I visited Star City again in 1994, but it proved to be the last time I was able to stay with my friends the Gaydukovs. Sadly, Valeriya Gaydukova fell seriously ill later that year and passed away in January of 1995. Gaydukov could obviously not host me himself anymore, and asked the Yesins if they'd be willing to let me stay with them when I next visited; they said they would be happy to.

Between 1998 and 2007 I would visit Star City with fellow space sleuth Rex Hall and, on a number of occasions, also with Neil Da Costa. Putting up two or even three people was asking too much of the Yesins, and whenever I wasn't on my own we would stay at the Orbita Hotel in Star City. It wasn't a hotel where one could simply book a room; it was where more or less official guests of the GCTC or people in Star City could stay.

During all these visits, I still tried to focus on interviewing unflown cosmonauts. As time went by, inevitably the older generation was losing members and I wanted to try to at least have their voices on tape and record their own stories, pretty much like an Oral History project at NASA. With the help of the Yesins, I also managed to get in touch with unflown members of other cosmonaut groups, such as NPO Energiya, the IMBP and the former Chelomey design bureau.

A special group were the Soviet space shuttle cosmonauts – in particular members of the so-called 'Wolf Pack', a group of test pilots headed by Igor Volk (his surname means wolf in Russian) that came from the Ministry of Aviation Industry's Gromov Flight research Centre in the city of Zhukovskiy, outside Moscow. As always, I had started my search for information by writing letters. Magomed Tolboyev had given a follow-up letter to Lida Shkorkina, and asked her to answer my questions. She knew all the group members personally, as well as the widows of those who were dead. A teacher, Lida was fluent in English and our on-going correspondence has resulted in me visiting regularly. She arranged meetings with the Buran cosmonauts, not only the ones from the Gromov Flight Research Centre but also those from the air force Flight Test Centre in Akhtubinsk now working for the aircraft industry in Zhukovskiy. Her help was indispensable in writing *Energiya-Buran – The Soviet Space Shuttle* [12].

BAYKONUR AT LAST!

In February 1997, I got the chance to join a group of German journalists travelling to Baykonur for the launch of Reinhold Ewald, a German cosmonaut who would fly a short mission to Mir under the Mir97 programme of the German space agency DLR. This was, of course, a great opportunity to see the city and launch site first-hand. To my surprise, we were not restricted in any way when we wanted to take photographs. Unfortunately, the trip had been arranged only a few days in advance, so there was not much time to sit down and prepare, but I did the best I could.

7.9: The Buran shuttle gathering dust in Baykonur.

Due to my contacts with Lida, and the members of the Wolf Pack, I already had a special interest in the Buran programme and I was thrilled when we were taken to the building in which the orbiter (which had made the single, unmanned spaceflight in 1988) was stored. There were three central core stages for the Energiya rocket and lots of strap-on boosters. In another hall was the orbiter itself, covered in a layer of dust. Our guides had no problems with me taking pictures, but initially didn't want me to climb the scaffolding around the Energiya and the Buran. I thought that they were opposed to me taking close-up pictures, but it turned out that they were simply worried I might fall off and hurt myself. Once I had assured them that they needn't worry, they let me do whatever I wanted to do.

Doors closing again?

In December 2011 I had another chance to visit Baykonur but found that things were more restricted this time. Since my last visit, the roof of the hall housing Buran had collapsed, so I wanted to try and photograph the rubble and any recognisable parts of what was left of the orbiter. However, there was no chance of anyone letting me into the building.

Another place I wanted to visit was the cemetery. In October 1960 the infamous Nedelin disaster occurred. A fully fuelled two-stage R-16 ICBM had exploded whilst being worked on, killing dozens of technicians as well as Chief Marshall of Artillery Mitrofan Nedelin. Although the fact of the disaster was an official secret until 1989,

Jim Oberg had done research into the event and published it in his book *Red Star in Orbit* [13]. Officially, 74 people were killed in the explosion but there were stories that many more people had died, possibly as many as 120. What I wanted to do was go to the cemetery and try find graves of people who had died on or shortly after 24 October of that year. If they were military and relatively young, they just might be additional victims. In the city itself, a monument had been erected to commemorate the personnel of the Scientific Research Test Range #5 (NIIP-5) who had died. Two plaques on the monument list the names of 54 of the 55 victims. One is missing (Lt-Col. Sakunov) but there's no clear reason why – after all, his name is on the list of casualties and is also mentioned on the monument at the site of the explosion. Urns with the ashes of 53 people are buried beneath the monument. Besides Sakunov, a second victim is missing: Capt. Kalabushkin. Why these two aren't buried there and why Sakunov's name is not mentioned at all is unknown. There is still work for the space sleuths!

Photography limited

I had decided to attempt to find out, but was told that I could not visit the cemetery. The reason given was that it was outside the city. My suggestion to have a security officer accompany me came to nothing. All I could do to enlarge our knowledge of the explosion was to examine the displays in the local museum. Most foreigners pass quickly through these rooms filled with portraits and other displays of the early years of the base, but they are historical gems. With a little luck, the staff will allow you to photograph the photographs that are on display.

Another event showing that photography was limited compared to my first visit occurred during the erection of the Soyuz rocket which was to be launched to the ISS. Unlike my first visit, I was restricted to only a small area, and once the rocket was in position and we walked back to the bus a number of our group were told not to take any pictures of the buildings in the industrial area over three kilometres away! It was of course ludicrous, because several days later we were taken there and allowed to photograph anything we wished!

Over the years, for me space sleuthing has proven to be a combination of knowing what to look for, realising you're coming across something remarkable when you do, and being in the right place at the right time: today they'll let you take all the pictures you want, tomorrow taking a picture can be prohibited. There's no rationale behind it whatsoever, which may make it difficult to decide what to do.

A hospital 'emergency'

In 2008, I had been told that a Buran crew cabin was sitting near the entrance of a hospital in Moscow. During my trip that year, I decided to go to the hospital and ask for permission to enter the grounds and take photographs. I couldn't imagine that this might pose a problem. After all, it was twenty-year-old obsolete technology.

The hospital security guard said he could not make a decision, and so his superior was called. That guy also would not make a decision, so he called *his* superior, who denied me permission. No reason was given. It was "Nyet", and that was that. There was nothing else to do but leave the hospital. On returning to the street, I wanted to photograph the cabin even though it was now almost 100 metres away and partially blocked by the fence that surrounded the hospital. It was annoying, but almost twenty years after the fall of communism there was no telling what the Russians would or wouldn't allow. I figured that every so often one would find a 'nobody' who was finally able to be a 'somebody'. Whatever the case, I never took the picture because a woman emerged from the building that I'd just left screaming as if she was being physically attacked.

A minute later I was stopped by an armed guard who ordered me to go with him, back to the hospital. Two more officers with walkie-talkies met us halfway, and once inside I was marched to the office of what appeared to be the head of security. In the meantime, policemen with machine guns had also arrived and my papers and camera were carefully studied. Of course copies of my passport and visa were made, as was always done when I visited certain space research facilities. There had to be hundreds of them in all sorts of files by now. It didn't bother me as much as the greedy look in the security guy's eyes when he was holding my camera. It wasn't too expensive but had a memory card that already contained almost 200 pictures that I'd taken earlier in the week in Star City and the GCTC. I was getting nervous that he might confiscate – steal – my camera and that I'd lose those photos.

In the end, it was the police officers that made it clear to him that they didn't really see any wrongdoing, and they wouldn't make a report. After all, I'd wanted to shoot the pictures from outside the hospital grounds. The security guy said that even if the police didn't file a report, he would. I was released. I don't know if a report was filed, but I never heard anything about the incident.

Sleuthing in the modern era

When I first started researching the Soviet manned space programme I was gathering every small bit of information trying to get the general picture. At the time, very little was published in official sources and it was clear that there was a whole hidden story waiting to be uncovered. Since the late 1980s and early 1990s, things have changed a lot. The Russians themselves have a space magazine, *Novosti Kosmonavtiki,* which is considered by many to be the best in the world. It regularly contains great articles on the history of the space programme. The space sleuthing of twenty years or more ago has mostly gone. Now much can be found in the excellent books that are published in Russia or on the Internet.

It will always be worthwhile taking your camera wherever you go, no matter how often you've been there. In 1999 Rex Hall and I were shown around the large hall in which nowadays the Russian ISS training modules are located. At that time there was a huge curtain dividing the hall. One part housed training mockups for the Spektr and Priroda modules of the Mir space station. While Rex was being told about these two modules, I had the opportunity to peek behind the curtain. I was stunned when I saw two Buran simulators: a fixed base and a motion base one.

Alongside, there was also a very basic mockup of the TKS descent module. This

7.10: A Buran flight simulator in 1999, literally hidden behind a curtain.

7.11: Bert Vis in front of what is left of the Buran simulator after it was dismantled and put on display in the ISS training hall (April 2011).

was the first time (and the last, for that matter) that I ever saw any training hardware for the transport ship that Chelomey's bureau designed for their military Almaz space stations.

Some people say that the days of space sleuthing are over. I don't agree. The way in which research is conducted may have changed, but in spite of everything that has been published since the early days of *glasnost*, there's a lot still awaiting discovery. Besides, after the lifting of secrecy in Russia there's a new target. The secrecy which surrounds the Chinese manned space programme is different from that of the former Soviet Union, but there are similarities. At ASE congresses Chinese astronauts refuse to identify their unflown colleagues. No reason is given. And so, the game continues.

References

1. Letter to Vis from the GCTC Information Group, 2 September 1987.
2. GCTC Information Group letter, 21 July 1988.
3. *Spaceflight*, February 1989.
4. GCTC Information Group letter, 4 October 1988.
5. Letter from Anatoli Voronov, 13 October 1989.
6. Undated letter from Anatoli Voronov, received on 3 August 1990.
7. Gordon Hooper, *The Soviet Cosmonaut Team*, GRH Publications 1986/1990.
8. *Spaceflight News*, August 1990, September 1990 & February 1991.
9. Michael Cassutt, *Who's Who in Space*, G. K. Hall & Co. 1986/1992/1996.
10. Interview with Valeri Yazdovskiy, Star City 14 April 2000.
11. *Russia's Cosmonauts, Inside the Yuri Gagarin Training Centre* by Rex Hall, David Shayler and Bert Vis, Springer-Praxis Publishing 2005.
12. *Energiya-Buran – The Soviet Space Shuttle*, by Bart Hendrickx & Bert Vis, Springer-Praxis Publishing 2007.
13. *Red Star in Orbit*, by James E. Oberg, Random House 1981.

8

Russian-language sleuthing

by Bart Hendrickx

Having been born just too late to witness the exciting events in space in the 1960s and early 1970s, my fascination with the Soviet space programme did not develop as naturally as it probably did for the "first-generation" sleuths. In fact, my first passion was astronomy, and as a child I spent many nights gazing at the sky through a small amateur telescope. I became interested in spaceflight as a teenager in the late 1970s through the Voyager missions to the outer planets. With the Space Shuttle still a long way away from its maiden flight, the only country launching piloted spacecraft at the time was the Soviet Union. Up in orbit then was the highly successful Salyut 6 space station and I began following the final missions to the station via the Western media.

Back issues of *Space World* and *Spaceflight* available in the library of a local observatory sparked an interest in the earlier history of Soviet spaceflight. What especially whetted my appetite for the sleuthing business was James Oberg's 1981 book *Red Star in Orbit*, which I even partially translated into Dutch (my mother tongue) as a thesis project for school. I was amazed by the amount of evidence for the Soviet manned lunar project, the military Salyuts, and those 'missing cosmonauts'. I became fascinated by the mystery surrounding so many aspects of the programme and the challenge of getting to the bottom of things. It was the beginning of a passion that has never faded since.

DISCOVERING RADIO MOSCOW

In 1980, while scanning the airwaves with my shortwave receiver, I accidentally stumbled on Radio Moscow World Service, which broadcast in English 24 hours per day and had almost daily updates on activities aboard Salyut. From then on, listening to the Radio Moscow newscasts became an almost daily routine. I recorded them and made transcripts of space-related news items. Not that these contained any shocking revelations, but in the pre-Internet era this was simply the most convenient

8.1: Bart Hendrickx is a qualified Russian-language translator.

way of keeping up to date on the Soviet space programme. The Western media didn't pay much attention to these Salyut missions unless something spectacular happened, and the mission reports in *Spaceflight* didn't appear until months later.

Meanwhile, mostly because of my enthusiasm for spaceflight, I decided to take up Russian. This allowed me to follow the domestic Russian radio broadcasts available on shortwave, particularly the first and second channels of Vsesoyuznoye Radio (All-Union Radio). The second channel – also called Mayak (Beacon) – carried interviews with Salyut crews by its space correspondent Pyotr Pelekhov. Of course there was no live coverage of launches, dockings, spacewalks or landings but recorded live reports were usually replayed after the event.

Although Salyuts 6 and 7 were civilian space stations, it was obvious the Russians weren't telling us everything. For instance, in April 1981 they launched Kosmos 1267. This was announced as a routine satellite in this series, but eventually it headed for Salyut 6 – arriving there two months later. We now know that it was a transport vehicle built by the Chelomey bureau that was originally designed to fly to the military Salyuts (Almaz), but the Russians were extremely secretive about it and nobody had a clue at the time as to what it was. I remember becoming obsessed with this spacecraft, spending many hours trying to figure out what it might look like and what its role might be. Although none of my fancy theories turned out to be correct, these were my first, modest attempts at doing some sleuthing of my own.

Launch pad explosion kept secret

Back then failures were also covered up. The most extreme example of this was the explosion of a Soyuz rocket on the launch pad on 26 September 1983. Fortunately, ground controllers managed to activate the emergency escape system to save the lives of cosmonauts Vladimir Titov and Gennadi Strekalov. I remember hearing a report on the American Forces Network the following day that spy satellites had seen a large plume of smoke over the Soviet Union. Finally, on 1 October the *Washington Post* quoted US intelligence sources as saying that a manned Soyuz had exploded on the pad and the crew had survived. Meanwhile, Radio Moscow and the domestic service continued their routine coverage of the ongoing Salyut 7 mission, giving no clue that anything had gone wrong. Vladimir Lyakhov and Aleksandr Aleksandrov (who were to have been replaced by Titov and Strekalov) did acknowledge the accident at their post-landing news conference on 12 December. But the news was only intended for a Western audience – it was reported by the World Service, but the Soviet public was not informed, even though Soviet reporters must have attended the news conference.

GLASNOST ON THE AIRWAVES

Things began to change after Mikhail Gorbachev became General Secretary of the Communist Party in March 1985 and launched the policy of *glasnost* or "openness". Gorbachev's rise to power came just about a month after the Salyut 7 space station had broken down while orbiting the Earth unmanned. As I remember it, the first sign of *glasnost* in the space programme came on 4 May, when Radio Moscow reported that cosmonauts Vladimir Dzhanibekov and Viktor Savinykh had started training at Star City "under a new programme", which was an implicit admission that they would be the next crew to fly to Salyut 7.

It was quite unprecedented for an all-Soviet crew to be announced in advance, but that is just about as far as *glasnost* went for the time being. When the two men were launched aboard Soyuz-T 13 on 6 June, there was no word that the crew was facing the daunting task of docking with and then repairing the derelict space station. Even as the cosmonauts were in the midst of performing one of the most heroic missions in the history of spaceflight, the Soviet media behaved as if it was business as usual by devoting more attention to VeGa 1's arrival at Venus. It was not until the publication in *Pravda* of an article by Konstantin Feoktistov in August that some of the hardships suffered by the men during the risky rescue operation came to light.

Glasnost was again put to the test later that same year when cosmonaut Vladimir Vasyutin fell ill aboard Salyut 7, forcing the early return of the Soyuz-T 14 crew in November. The only indication that something was not right in the final weeks of the mission was the fact that there were fewer mission updates on Radio Moscow and no reference was made to the health of the crew. Although Radio Moscow admitted that the crew had returned prematurely due to health problems experienced by Vasyutin, no further details were forthcoming. Again it was *Pravda* that released at least some details about the situation in late December, when it published excerpts from the on-board diary kept by Vasyutin's crewmate Viktor Savinykh.

Live launch coverage

A major breakthrough in the policy of openness occurred on 12 March 1986, when Soviet radio announced that the crew of Soyuz-T 15 would be Leonid Kizim and Vladimir Solovyov, and promised live radio and TV coverage of their launch the next day. This was the first time that a Soviet manned space launch with no international involvement was broadcast live. I had no access to Soviet television, so I followed the launch in real-time on both Radio Moscow World Service and the domestic service. As would become standard procedure, the domestic service played the popular cosmonaut song 'Trava u doma' (Grass of Home) with the air-to-ground communications link in the background as the Soyuz soared into space, probably to make it easier to interrupt the broadcast if things went wrong. On the World Service veteran science correspondent Boris Belitsky provided running commentary on events until about two minutes into the flight. Again, Western listeners were told more than the Soviet audience. During the entire 40-minute broadcast on the domestic service, the reporters did not mention which space station the crew was headed for, saying only they were expected to work both on Mir and Salyut 7. Only later in the day were Russian listeners informed that the Soyuz was heading for Mir. No secret of this had been made in the World Service live broadcast and in newscasts preceding the launch.

An uncomfortable radio silence

The limits of *glasnost* were very much in evidence in early September 1988 when the Soviet-Afghan crew (Vladimir Lyakhov and Abdul Mohmand) was set to return to Earth from a visit to the Mir space station aboard Soyuz-TM 5. I got up early on the morning of 6 September to follow Mayak's live coverage of the landing, which was scheduled for 2.15 GMT. In an update at 1.45 GMT, Mayak's reporter at Mission Control quoted Lyakhov as saying "everything was going according to plan". This was about fifteen minutes after the scheduled deorbit burn. The next live update was promised for 2.00 GMT, but to my surprise Mayak did not interrupt its regularly scheduled programming. The scheduled landing time went by without any word on the fate of the crew. Since the last update had created the impression that the deorbit burn had gone smoothly, I naturally feared the worst. The whole situation conjured up images in my mind of the Soyuz 1 and Soyuz 11 disasters many years before.

The eerie silence continued until 3.30 GMT, when Mayak broadcast a terse TASS communiqué saying the landing had been delayed for technical reasons. Only later in the day did it emerge that the deorbit burn had failed and that Lyakhov and Mohmand were stranded in space for at least the next 24 hours. The following morning Mayak was again silent as events unfolded, with no live commentary from Mission Control. Confirmation of the safe landing did not come until 1.30 GMT, forty minutes after it occurred, and it was only several hours later that the recorded live landing report was replayed. Subsequent reports were more frank about the nature of the mishap.

Eavesdropping on the cosmonauts

Another way of keeping track of the Mir missions was by listening to the cosmonauts themselves. I became aware of this possibility when Dutch radio amateur Chris van den Berg began providing updates on the Mir missions in a weekly radio programme by playing back and analysing air-to-ground conversations that he had picked up. All I needed to do was to buy a small handheld FM scanner and tune in to 143.625 MHz as Mir flew over. Since they could also be used to listen to the police, such scanners were illegal in Belgium but it was easy to pop over the border to Holland (where they were openly for sale) and purchase one.

Eavesdroppers in Western Europe had the good fortune of being able to listen to the cosmonauts when the station came within range of Mission Control near Moscow. If something interesting had happened during the "loss of signal" period, we would be the first to find out. Of course, even during the best passes the station was over the horizon no longer than 10 minutes and reception was not always clear throughout the pass. Most of the conversations were mundane (or so technical you didn't know what

8.2: Chris van den Berg seated at the radio of a Soviet tracking ship.

they were talking about) but sometimes they were fascinating – in particular during docking operations and spacewalks. For instance, after the Progress-M 7 cargo ship failed to dock with Mir twice in 1991, I overheard the crew talking about activating their Soyuz spacecraft, an indication that something interesting was about to happen. Only afterwards did the Russians announce that the crew had flown their Soyuz over to the other docking port to investigate the fault with the Kurs rendezvous antenna of the station that had foiled the Progress docking attempts.

Lack of time made it impossible for me to regularly listen to the cosmonauts, but enthusiasts like Chris van den Berg were able to glean lots of interesting information that was not available from the normal Russian channels, especially after the collision of a Progress ship with Mir in 1997.

ENERGIYA ROCKET MAKES ITS DEBUT

Another spectacular demonstration of *glasnost* came on 14 May 1987, when Soviet radio and other media reported that final preparations were underway at Baykonur for the launch of a new-generation booster "capable of taking into orbit reusable craft and large modules for research and Earth resource studies". This was the first public revelation of the Energiya-Buran programme. I vividly remember hearing the report on the domestic service and it is hard to describe the excitement that I felt at the time. For years we had been reading reports in *Aviation Week & Space Technology* and the annual US Defense Department publication *Soviet Military Power* about the Soviet heavy-lift rocket and shuttle, but still, it was almost surreal to suddenly hear that they really existed.

With the Soyuz launches now being carried live, I expected to see live coverage of the rocket's maiden flight, so it was a major disappointment when Radio Moscow did not announce the launch until after it had occurred on 15 May. With the failures of the N-1 still fresh in their memories, officials clearly did not want to take the risk of a worldwide audience witnessing a spectacular launch failure. Although Energiya itself performed flawlessly, the payload failed to place itself into orbit due to a navigation error. Had the launch occurred in the pre-*glasnost* days, it would probably have been covered up for that very reason, but TASS openly admitted that the payload (only referred to as a "dummy satellite") had re-entered shortly after launch. In fact, this payload, secretly called Polyus or Skif-DM, was the prototype for a space-based laser battle station developed under the Soviet equivalent of the American SDI programme. If this had made it into orbit, it would have been interesting to see how the Russians handled the public release of information on its mission, as they had heavily criticised America's 'Star Wars' effort.

Of course, I couldn't wait to see the first pictures of the big rocket, as no images had been released prior to launch. On 16 May Western television stations replayed some of the footage of the launch preparations and the lift off aired earlier in the day by Soviet television. Interestingly, the payload was carried strapped to one side of the vehicle, rather than on top. Most of the views were of the side opposite the payload, but one distant side-view included the large, black pencil-shaped object. Somehow, most Western observers do not seem to have taken note of this at the time, possibly because this particular perspective was not widely featured in the Western bulletins. The payload remained out of view in pictures of Energiya released subsequently, and it would be many more years before pictures of Polyus became available.

The Soviet shuttle

A year later final preparations were in full swing for the long-awaited maiden flight of the Soviet space shuttle Buran. Although the existence of this programme had been acknowledged by the Russians in the spring of 1987, Soviet space officials adopted a carefully controlled publicity concerning their plans, revealing virtually nothing about the status of launch preparations. The handful of statements in the following months came from officials of Glavkosmos. Although portrayed in the West at that time as being the Soviet equivalent of NASA, Glavkosmos was actually the international-relations arm of the Ministry of General Machine Building – the top-secret ministry that oversaw the space industry. In March 1988 Glavkosmos chief Aleksandr Dunayev promised the first Soviet shuttle launch would be shown live on television.

In the West, speculation was rife that the Russians would try to upstage the return to flight of the American shuttle after the January 1986 Challenger accident. In the event, Discovery blasted off from Cape Canaveral in Florida on 29 September 1988 with its Soviet counterpart still on the ground at Baykonur. Having been through the excitement of the first shuttle launch in two and a half years earlier in the day, space buffs were in for yet another treat. Intent on stealing at least some of the thunder from NASA's success, the Russians took advantage of the occasion to finally unveil their shuttle to the world. Just hours after Discovery's launch, the Soviet evening TV news programme *Vremya* opened by showing the first-ever publicly released picture of the Energiya-Buran stack on the launch pad. Interestingly, the name Buran on the side of the vehicle had been retouched so as not to be visible to the television audience, even though the name had leaked to the West several years earlier. It was not stated when the launch was to occur, but *Vremya* did say that the mission would be unmanned.

The Soviet shuttle receded into the background until 23 October, when *Vremya* showed the first footage of Energiya-Buran, including spectacular shots of the roll-out to the pad. Finally, on 26 October Soviet media reported the launch would occur on 29 October at 3.23 GMT. However, in a major disappointment to space buffs all over the world, on the eve of the launch Soviet officials retreated from their earlier promise to provide live television coverage of the launch, and were now planning to show it 35 minutes after the event.

I was up and awake early in the morning of 29 October in anticipation of what to me personally was the most exciting space event since STS-1 in April 1981. I was glued to my shortwave receiver as the planned launch time of 3.23 GMT passed without a report. After 45 minutes, in a major anti-climax, Mayak announced that the mission had been postponed for four hours. It was later rescheduled for 15 November at 3.00 GMT.

8.3: The first official TASS image of Soviet shuttle Buran.

On that day, both Radio Moscow World Service and the Soviet domestic Mayak radio station reported the launch at the very beginning of their 3.00 GMT newscasts. The World Service even optimistically stated that Buran had been placed into orbit, although orbital insertion was still at least ten minutes away. At 4.10 GMT Mayak broadcast a recorded live report of the launch from its reporters at Baykonur and in Mission Control in Kaliningrad near Moscow.

There was no live coverage of the landing either, but both Radio Moscow World Service and Mayak reported the safe return only minutes after touchdown. Half an hour later, Mayak aired their recorded live report. One of the reporters couldn't contain his enthusiasm, shouting "Hurray!" and "Victory!" as Buran rolled to a stop on the runway. Little did anyone know at the time that Buran would never return to space.

NEW WINDOW ON SPACE HISTORY

Not only did *glasnost* allow more open coverage of current Soviet space activities, it also gave journalists an opportunity to delve into the past and uncover some of the space programme's less glamorous episodes. While continuing to follow current events, my interest gradually shifted to this earlier history. These revelations opened a new chapter in the sleuthing business. It became a matter of analysing the newly released Russian information and combining that with information from Western sources, then filling in the blanks by applying educated speculation.

Most of the early revelations came in regular newspapers such as *Pravda, Izvestia* and *Krasnaya zvezda* as well as popular magazines. I didn't subscribe to those, so I usually found out about new revelations from the Western press or space magazines such as *Spaceflight*. Because these normally carried only summaries of the Russian publications, the next step was to track down the original articles – which was not always easy in those days. Fortunately, I was able to find a lot of the material I was looking for in a bookshop in Brussels specialising in Russian books and magazines. Of course, it was impossible to find every original Russian article and what came in very handy in those days were English translations of Soviet press articles regularly published by the US Foreign Broadcast Information Service in a publication called *JPRS Report/USSR: Space*.

The first major revelations centred on the Soviet cosmonaut team. In 1986 veteran space journalist Yaroslav Golovanov was allowed to publish the names of the eight unflown cosmonauts of the original "Gagarin" team" recruited in 1960, and in 1987 the women's magazine *Rabotnitsa* unveiled the names of Valentina Tereshkova's four fellow trainees [1]. Also in 1987, the popular weekly *Ogonyok* printed an article on Sergei Korolev's prison camp ordeals, and two years later another one on the "Nedelin disaster" which had killed dozens of officials and launch pad workers as a rocket blew up on the pad at Baykonur in October 1960 [2]. For years there had been speculation in the West about the accident (nicely summarised in Oberg's *Red Star in Orbit*). Having occurred during a Mars launch window, it had long been believed the disaster involved a rocket carrying a Mars probe but *Ogonyok* now disclosed that the catastrophe had happened during preparations for the first test launch of the Yangel bureau's R-16 ICBM.

Unveiling the manned lunar project

By 1989 *glasnost* had progressed far enough to enable the Russians to lift the veil on one of the most closely guarded secrets of the Soviet space programme – the ill-fated manned lunar project of the 1960s and 1970s. Prior to Apollo 8 and Apollo 11, Soviet cosmonauts and officials regularly pointed out that the USSR was indeed in the Moon Race without setting specific timelines or giving away any technical details about the spacecraft or boosters. Of course, after the triumph of Apollo 11 the Russians began denying they had been in the race, and it is quite amazing that this was believed by so many in the West.

Sleuths had much more to go on than just the Soviet statements prior to Apollo 8. The evidence was all over the place. There were leaked US intelligence reports on the launch failures of a giant booster in the same class of the Saturn V, mysterious Earth-orbital flights under the guise of the "Kosmos" programme that carried tell-tale signs of being lunar hardware tests, and clues that the Zond 4, 5, 6, 7 and 8 vehicles were stripped-down Soyuz spacecraft designed to fly cosmonauts around the Moon. One didn't even have to read between the lines to find confirmation that the Zonds were part of the manned programme. It was there in black and white. After the landing of Zond 6 in November 1968 *Pravda* specifically stated that Zond 4, 5 and 6 had been launched "to obtain important data on the functioning of the design, the on-board systems and equipment of a piloted spaceship for flights to the Moon". A similar phrase was included in the entry for Zond published in the original 1969 edition of the *Soviet Encyclopedia of Space Flight*. Predictably, the phrase was deleted from the second edition published the following year – after the US had won the Moon Race [3].

Even after the Soviets began vehemently denying their involvement in the Moon Race, information occasionally slipped through. For instance, in 1977 a Soviet book about launch sites published a drawing of the rotatable N-1 launch pad gantry [4]. Although it didn't mention what the gantry had been intended for, Western sleuths (notably Charles Vick) readily identified it as the gantry of the lunar rocket, allowing them to deduce the size of the vehicle. In August 1981, when the Kosmos 434 spacecraft (a test version of the Soviet lunar module launched into Earth orbit for manoeuvring tests in 1971) was about to re-enter the Earth's atmosphere, the Soviets, to counter speculation that it carried a nuclear reactor, described it as a "lunar cabin", a term usually used for a manned lunar vehicle.

First admission of the truth

The first revelations of the manned lunar programme were quite illustrative of how *glasnost* worked in the early days. Almost teasingly, the information was released in dribs and drabs, often raising more questions than it answered. Whether by design or not, some of this information proved to be wrong.

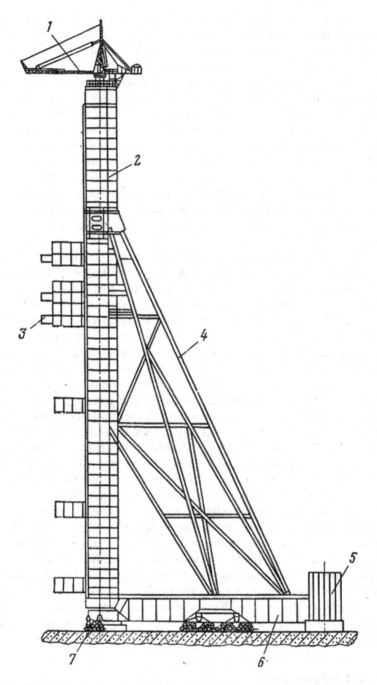

8.4: The N-1 launch gantry drawing inexplicably used in a Soviet book.

To the best of my knowledge, the first acknowledgment of the Soviet involvement in the Moon Race came in July 1989, when a magazine called *Poisk* published some heavily edited excerpts from the diaries of General Nikolai Kamanin, the Air Force official in charge of cosmonaut training from 1960 to 1971 [5]. These included some notes on both the manned circumlunar and lunar landing programmes, which were now publicly identified as L-1 and N-1/L-3 respectively. In an excerpt for late November 1968, Kamanin admitted that the Soviet Union was not yet ready to send an L-1 vehicle around the Moon. For years, Western analysts had thought the Soviets attempted to stage a last-ditch effort in early December 1968 to beat America in the Moon Race only to have their manned Zond mission scrubbed at the last moment. Kamanin's notes cast serious doubt on this claim. Other fragments from the diaries dealt with problems facing the manned lunar landing project, with the entry for 4 July 1969 verifying the long-suspected N-1 launch failure the previous day.

Having appeared in a rather obscure publication, these surprising revelations went unnoticed in the West at the time. The first disclosures about the lunar programme that did catch the attention of the Western media came in a biography of cosmonaut Valery Bykovsky [6]. In early August 1989, the TASS news agency made use of the imminent publication of the book to officially admit the Soviet Union's involvement in the Moon Race, attributing the country's defeat in the race at least partially to the death of Sergei Korolev in 1966. When the book was released, the section concerning the lunar programme was disappointingly meagre at just three paragraphs. Bykovsky was assigned to the L-1 circumlunar programme along with Leonov, Rukavishnikov and Kubasov. After two years, the programme was terminated, but the book did not elaborate on the reasons for that decision.

The first article specifically devoted to the subject appeared on 18 August 1989 in *Izvestia* [7]. Despite *glasnost*, it took the newspaper's science editor Sergei Leskov some effort to obtain permission for its publication, with the final stamp of approval coming from no less a person than Oleg Shishkin – the Minister of General Machine Building [8]. The article did not mention the circumlunar effort and focused instead on the history of the N-1 rocket. Its development had been approved by a government decree in 1960, but there followed a dispute between Korolev and Glushko over the choice of propellants, and this resulted in engine development being assigned to the relatively inexperienced Kuznetsov bureau. One big surprise was that there had been four rather than three N-1 launch attempts as earlier speculated in the West. It turned out that US intelligence had missed the first one on 21 February 1969. The dates for the second and third launch attempts were incorrectly given by *Izvestia* as 3 July 1970 (it was 3 July 1969) and 27 July 1971 (it was 27 June 1971). Absent from the article were any details on the lunar spacecraft themselves and the exact mission scenario, apart from the fact that two cosmonauts would have been sent to the Moon with one landing on the surface.

Former chief designer speaks

More excerpts from Kamanin's diaries were published in the *Sovetskaya Rossiya* newspaper on 11 October 1989, revealing the existence of a competitor to the N-1, namely the Chelomey bureau's UR-700 [9]. More information came to light nine days later when *Pravda* published an interview with former chief designer Vasily Mishin [10]. Mishin said the L-1 programme had been ready to send cosmonauts around the Moon, but after the Apollo 11 landing there was no propaganda value in such a mission. He also revealed that the N-1 had thirty engines in its first stage and used LOX/kerosene in all its stages – news that contradicted Western speculation that at least some of the stages used hypergolic propellants. One of the major handicaps had been the inability to test-fire the complete first stage of the rocket on the ground (as was done for its Saturn V counterpart). Remarkably, Mishin gave the same incorrect dates for the second and third N-1 launch as *Izvestia* had two months earlier. He stressed that if the N-1 programme had continued, the Soviet Union would by now have had a lunar base.

American engineers shown lunar hardware

What still remained a mystery at this point was the exact N-1/L-3 mission profile and the configuration of the lunar payload. Finally, in November 1989 a group of US aerospace engineers from the Massachusetts Institute of Technology were shown the LK lunar lander and elements of the LOK lunar orbiter during a tour of the Moscow

8.5: Vasily Mishin, former chief designer.

Aviation Institute (MAI). As it turned out, the LK had a descent and ascent stage like the US lunar module, and the LOK was essentially a modified Soyuz spacecraft. The MIT team was allowed to take pictures, some of which were published in the Western press in the following weeks [11]. Upon their return, they claimed to have been told by the Russians that the Soviet lunar landing mission would have used a dual-launch profile. The N-1 would have placed the unmanned lunar lander into Earth orbit, and the two-man LOK crew would have followed on a Proton. After the LOK docked with the lunar module and the upper stage of the N-1, this stage would have propelled the complex out to the Moon. Once in lunar orbit, one of the cosmonauts would have transferred externally to the lunar lander to begin the descent to the surface.

We now know, of course, that the N-1/L-3 mission profile envisaged a single launch by the N-1, raising the issue of why the US delegation was wrongly informed at the time. One possibility is that the erroneous information resulted from mistranslation or misinterpretation – not uncommon when technical information is exchanged through interpreters. Another is that the Americans were deliberately misinformed in order to keep observers in the dark about the actual mission scenario, although one wonders whether there was any reason to do so.

A third possibility is that the Russians guiding the MIT team around didn't know the real mission profile, and were simply echoing Western speculation. For years the most commonly held belief amongst sleuths had been that the lunar landing mission would have involved the launch of a Saturn V-class booster and a Proton. Some even thought a double countdown had been conducted in early July 1969 for a high Earth orbital manned test mission of the lunar vehicles and that the manned Proton launch had been cancelled only following the explosion of the N-1. It is possible that in the absence of better information, the Soviet guides of the MIT delegation believed this scenario to be true. The person leading the tour was MAI Professor Oleg Alifanov, who had never been personally involved in the lunar programme and may not have been privy to all its details (even though Mishin also taught at the Institute).

There were also cases of Western speculation about the lunar programme turning up in Russian publications after the collapse of the Soviet empire. For instance, in an article in 1993, the army newspaper *Krasnaya zvezda* published a strange drawing of a lunar landing craft that consisted of a beefed-up Soyuz attached to a Proton's third stage [12]. This was presented as a genuine Soviet-era proposal but in actual fact it was a slightly changed version of a drawing that had appeared in a 1987 article in *JBIS* (*Journal of the British Interplanetary Society*) describing a possible "direct ascent" profile for the manned lunar landing mission [13].

Cosmonauts talk at last

Despite the first revelations about the lunar programme, not all cosmonauts who had been involved felt comfortable talking about it. When Steven Young of *Spaceflight* magazine asked cosmonauts about their participation in the programme during the Association of Space Explorers congress in Riyadh, Saudi Arabia, in November 1989 he got mixed reactions. While Oleg Makarov and Georgi Grechko

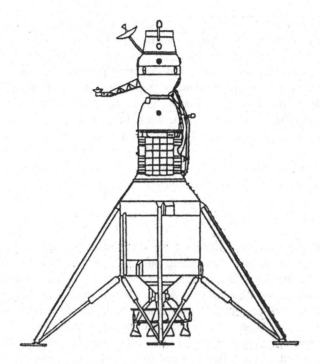

8.6: A Russian drawing copying incorrect Western speculation.

spoke candidly, Alexei Leonov (once slated to become the first Soviet cosmonaut to circle the Moon and eventually set foot on it) continued to deny strongly there had ever been a Soviet programme to land a man on the Moon or even to fly around it. This was despite the fact that he had been personally linked to the programme in the Bykovsky biography published only months earlier and also by Makarov at the very same conference [14].

Conflicting information

With the news from the MIT delegation reinforcing Western observers in their belief that they had been correct about the mission profile, new details released in the Soviet press contradicted the dual-launch scenario. On 13 January 1990, just days after the authoritative *Aviation Week & Space Technology* had published its account of the MIT team's visit to Moscow, *Krasnaya zvezda* published reminiscences by several veterans who had been personally involved [15]. One of them, TsNIIMash director Yuri Mozzhorin, succinctly described the launch and mission profile but made no mention whatsoever of a Proton rocket. Accompanying the article was a drawing of the N-1 that bore only a superficial resemblance to the true design, but was almost a carbon copy of a hypothetical drawing published in *JBIS* in 1985. Again, Mozzhorin repeated the same false dates for the second and third N-1 launches given in earlier publications.

Although *Krasnaya zvezda* was a widely read newspaper, the new information did not immediately reach the West; not even when the English translation of the article appeared in *JRS Report* three months later [16]. While it was not surprising that the general press ignored the technicalities of the lunar project, the fact that the article escaped the attention of Western aerospace reporters – and indeed the sleuths – was quite remarkable. To some extent this can probably be attributed to the absence of an efficient communications network between sleuths in the pre-Internet era. The article may well have been spotted by Russian amateur spaceflight historians, but they had no platform to share such news with a larger audience. As a result, for the better part of 1990 Western analysts were left with the impression that the Soviet lunar landing mission would have involved a dual launch.

Leonov breaks his silence

While the regular press had uncovered some of the basic details of the programme, it would clearly require the more specialised publications to satisfy the demands of the real history buffs. In the summer of 1990 the Air Force aerospace journal *Aviatsiya i kosmonavtika* carried an interview with Alexei Leonov, who finally spoke of training for the L-1 circumlunar project [17]. Presumably by lapse of memory, Leonov mixed up some details of the Zond flight history, and also mistakenly said there were only three N-1 launch attempts. However, he did correctly disclose that the final scheduled unmanned L-1 test flight (Zond 6) in November 1968 had ended with a crash-landing, a fact that had been successfully hidden from Western observers. This cast even more doubt on the veracity of the rumoured manned Zond attempt in early December 1968.

Mission profile revealed

The first detailed account of the project was finally published at the end of 1990 in the form of a monograph by Vasily Mishin in the monthly magazine *Kosmonavtika, astronomiya* (*Spaceflight, Astronomy*) [18]. After a long introduction on automatic lunar probes and the US Apollo programme, he focused on the Soviet manned lunar programme and described both the L-1 circumlunar programme and N-1/L-3 landing project. The article was accompanied by the first accurate drawing of the N-1 with its lunar payload, and included a diagram with captions outlining the mission profile.

But Mishin's account was not accurate in every respect. He created the impression that the N-1 project had been assigned to the Korolev bureau in late 1961 and that the Chelomey bureau had been put in charge of the circumlunar project at the same time, with the latter getting priority in 1962 until it was decided in mid-1964 to focus on a lunar landing. In actual fact, development of the N-1 was authorised by a government decree in June 1960 and the circumlunar and lunar landing programmes did not gain approval until a decree in August 1964. Apparently mentioning these secret decrees was still a sensitive issue. Mishin gave the date for the second N-1 launch correctly as 3 July 1969, but in an accompanying timetable it was again listed as July 1970. The date for the third launch was again wrongly given as 27 July 1971. After summing up the reasons for the failure of the programme, Mishin stressed that it ought not to have been terminated just because the Americans reached the Moon first. He revealed that his bureau had worked on a more advanced dual-launch mission that would have put three cosmonauts on the Moon for 14 days – the length of one period of daylight in the lunar cycle.

News of the Russian publication reached the Western aerospace press in February 1991 [19]. Meanwhile, *Spaceflight* had learned the correct N-1/L-3 mission profile from Mishin during an interview at the IAF congress in October 1990 and by featuring this in its January 1991 issue it became the first Western magazine to publish it [20]. Because of its historical significance, the complete Mishin monograph was translated into English in a dedicated issue of *JPRS Report* in late 1991 [21]. The basic facts and figures of the Soviet piloted lunar programme were now known, but it had taken almost 18 months for them to surface.

RUSSIAN SLEUTHS IN ACTION

Whilst the sleuthing business is usually associated with Western analysts, there was also a community of Soviet spaceflight enthusiasts trying to read between the lines of official Soviet statements and asking the same questions as their foreign counterparts. However, they were handicapped by difficulty in accessing Western publications and the inability to openly speculate on hidden aspects of their own programme.

Circumventing censorship

Like their Western colleagues, the Russian sleuths became adept at reading between the lines of Soviet statements or were "given a helping hand" by Soviet publications. For instance, when the *Soviet Encyclopedia of Space Flight* came out in 1969, it had a list of all satellite launches conducted to that point, and this included several mysterious launches in late 1962 and early 1963 that had not been announced at the time. In the columns for "launching nation" and "name of satellite" was written "no data". On the other hand the orbital elements *were* given, and by comparing those with the initial orbital parameters of deep space probes that made it out of Earth orbit, it was easy to identify them as stranded Soviet Venus, Mars and lunar probes.

One way to find out about Western speculation on the Soviet space programme was to listen to the shortwave broadcasts of radio stations like the Voice of America and BBC World Service. The Russian-language broadcasts of these stations were regularly jammed, but due to the high costs associated with operating the jamming transmitters this could not be done on a permanent basis. And the English-language broadcasts were rarely if ever jammed.

Western journals edited

Soviet citizens *were* allowed to subscribe to magazines such as *Spaceflight* and *JBIS*, but only through an organisation called 'Soyuzpechat'. This distributed poor-quality reprints to its subscribers, making sure that sensitive material was omitted. Articles discussing such things as Soviet military space projects, the manned lunar project and unflown cosmonauts were usually censored – at least until the mid-1980s. Moreover, the subscriptions were relatively expensive by Soviet standards. Original copies of these magazines were restricted to technical libraries of space-related organisations or other libraries with limited access. These limited-access libraries also had a weekly bulletin called *Raketno-kosmicheskaya tekhnika* (*Rocket and Space Technology*) with translated Western articles on spaceflight (but not on the Soviet space programme), which for many in the space community (even the chief designers) was the best way to keep up to speed on developments abroad. The TASS news agency also compiled digests of Western articles on the Soviet Union but these obviously had a very limited circulation. [22]

Sleuths publish research

As *glasnost* eased publishing restrictions in the USSR, some of the spaceflight enthusiasts were finally able to come out into the open with their own research. They had two obvious advantages over Western analysts – they weren't hampered by the language barrier and they were much closer to the prime sources of information.

While professional science journalists or space programme veterans wrote many of the early articles revealing the secrets of Soviet spaceflight, amateur spaceflight historians began to play an increasingly important role. One of the most prominent ones in the early days of *glasnost* was Vadim Molchanov. Employed at the gas works of the city of Tula, Molchanov had no professional connection with space whatsoever but had spent a lifetime collecting information about the Soviet cosmonaut team. He had become particularly interested in the cosmonaut trainees who for one reason or another had never made it into space and whose names had therefore usually been erased from history.

As far as circumstances in the Soviet days allowed, Molchanov corresponded with and interviewed both flown and unflown cosmonauts. He saw his years of meticulous research rewarded in late 1990 when *Kosmonavtika, astronomiya* published a lengthy article he had written on unflown cosmonauts (both Soviet and American) [23]. This was the first significant work by an independent space historian to be published in the Soviet Union. Although many of the names and facts presented by Molchanov had already been uncovered by Western sleuths such as Rex Hall and Gordon Hooper, it was the first time they appeared in a Soviet publication. Unfortunately, Molchanov died in 1996.

At the end of 1991 *Kosmonavtika, astronomiya* published another landmark issue – unfortunately the last in the series to come out. Written by Igor Afanasyev, a young engineer then working for TsAGI, it was completely dedicated to formerly classified Soviet manned space projects [24]. This was a real treasure trove and I just

SPACEFLIGHT

88905   КОСМИЧЕСКИЕ ПОЛЕТЫ № Т-7
(спейсфлайт)
По подписке 1972 г

Volume 14 No 7 July 1972

Published by The British Interplanetary Society

8.7: Edited Soviet reprint of *Spaceflight*.

8.8: Pioneering Russian sleuth Vadim Molchanov.

couldn't believe my eyes when I first saw this. I remember reading it in one go – completely oblivious to what was going on around me. It confirmed many of the rumours that had been circulating in the West for many years, but also refuted others and took the wraps off projects that we'd never suspected. It provided many more details on the Soviet manned lunar programme, not only on the L-1 and N-1/ L-3 projects, but also on Chelomey's alternative lunar plans (LK-1 and UR-700/LK-700). It was the first publication to reveal the existence of some early spaceplane projects (like PKA and Spiral), and gave significant details about the Almaz military space stations of the Chelomey design bureau and the heavy transport ships (TKS) that were originally intended as transports for Almaz. Ten years after having become fascinated by these mysterious vehicles, I was very excited to see them described in such detail for the first time. Also included was new information on the origins of Soyuz and the first-ever description of a lifting body alternative to Buran and the Zenit-launched Zarya vehicle. Being a milestone in the historiography of Soviet spaceflight, this issue was translated in English for a wider audience [25].

First Russian 'space magazine' published

Whilst *glasnost* permitted unprecedented coverage of Soviet space missions and the declassification of formerly secret projects, it also exposed the space programme to criticism, with more and more people openly expressing scepticism about its cost and purpose. Moreover, as the country became increasingly embroiled in political turmoil and was headed for disintegration, the Soviet media had more pressing things to discuss than spaceflight.

There were no dedicated popular spaceflight journals in the Soviet Union to fill the information void. *Aviatsiya i kosmonavtika* was mainly for aviation, while magazines such as *Zemlya i vselennaya* and *Kosmonavtika, astronomiya* had astronomy in their coverage. In the second half of the 1980s the Ministry of General Machine Building came close to publishing a magazine called *Kosmonavtika (Spaceflight)*, but when a new Minister (Oleg Shishkin) came to power in 1989 he shelved this plan [26].

Frustrated by these developments, a team of young spaceflight enthusiasts of the Moscow-based Videokosmos organisation decided to rectify this lack of coverage in the media. They had specialised in making documentaries, but in August 1991 three of its staff (Igor Marinin, Sergei Shamsutdinov and Olga Zhdanovich) launched a bi-weekly publication called *Novosti kosmonavtiki (Spaceflight News)* that covered on-going space projects in the Soviet Union and abroad. It began as a very modest six-page newsletter reproduced on a copy machine and distributed for free amongst a few dozen spaceflight enthusiasts, but it soon evolved into a fully-fledged subscription-based magazine with an increasing number of pages. Joining the team of writers in the following years were talented analysts such as Igor Lissov, Konstantin Lantratov, Maxim Tarasenko and Igor Afanasyev. All were young people in their twenties with engineering backgrounds, some of whom had worked or were still working in the aerospace industry.

By the end of 1997 Videokosmos was on the brink of bankruptcy, but thanks to sponsorship from a Russian computer company *Novosti kosmonavtiki* survived and under the auspices of the Russian Space Agency it became a colourful monthly with 5,000 subscribers, and is now the world's most informative space magazine for both Russian and non-Russian projects. As the name of the magazine implies, the main focus is current projects, but it has also devoted considerable attention to historical research. By the efforts of its own journalists and regular contributors, a great many unknown aspects of the Soviet space programme came to light, both in the manned and unmanned fields.

Novosti kosmonavtiki was the first to expose several formerly top-secret Soviet military space projects. Memorable articles in the 1990s covered the broad family of DS satellites of the Yangel bureau and the Yantar reconnaissance satellites, enabling many satellites launched under the Kosmos 'cover' to be positively identified [27]. The magazine also revealed the existence of a series of unflown Soyuz-based manned reconnaissance vehicles that were secretly developed in the 1960s by a branch of the Korolev bureau in Kuibyshev [28]. After the turn of the century, coverage of Russian military space projects (both ongoing and historical) diminished – apparently due to restrictions imposed by the Putin administration on the declassification of military projects. But the magazine has continued to publish articles on less sensitive aspects of the Soviet space programme, and it regularly runs interviews with former Soviet cosmonauts who often divulge totally unknown facts about their flights.

BREAKING THE LANGUAGE BARRIER

As the information floodgates in the Soviet Union opened, only a small part of the newly released information trickled to the West, largely due to the language barrier. As a result, some persistent myths about the Soviet programme managed to survive even though they had been debunked by Russian publications – the dual-launch lunar landing mission being a case in point.

Russian sleuths rarely published the results of their research in the West. One of the few exceptions was Timofey Prygichev, a researcher based in Saint Petersburg with a lifelong passion for spaceflight and an exquisite sense of detail and accuracy. Writing under the pseudonym Timothy Varfolomeyev, he penned a widely acclaimed series of articles for *Spaceflight* in the 1990s on the history of the R-7 rocket and its many derivatives [29].

To some extent the language problem in the early days of *glasnost* was overcome by the English translations of Russian press articles in *JPRS Report* and also by the outstanding *Soviet Year in Space* publications written by Nicholas Johnson, which were largely based on Soviet-Russian sources. But these publications tended to focus on current events rather than history. And some of the new material was emerging in obscure publications that were hard to get and weren't translated. For instance, some revealing material on the manned lunar programme appeared in the early 1990s in an in-house magazine of the Moscow Aviation Institute named *Propeller*, and it was only through my contacts at *Novosti kosmonavtiki* that I was able to get my hands on photocopies.

Russian books – new source of information

Also, more and more revelations were being made in books – although finding the interesting tidbits was often like seeking the proverbial needle in a haystack. One of the first times I recognised this was during a trip to the Soviet Union in 1990 when I bought a book on some of NPO Lavochkin's deep space probes [30]. Hidden deeply in one of the chapters were a few paragraphs about an ambitious Mars sample return mission planned by that bureau in the early-to-mid-1970s which had gobbled up so much money that the Lunokhod 3 lunar rover mission was cancelled. It was the first time the Mars sample return project had been mentioned anywhere in open literature.

Having noticed that newly revealed facts like this often did not penetrate to the West, it became something of a personal mission to use my knowledge of Russian to convey the newly published information to a Western audience. The first step was to find such information. On my regular trips to Moscow, I spent a fair part of my time scouring the bookshops for new books. My Russian contacts were also always very helpful in locating new space-related literature. In the course of the years I have also expanded my library of Soviet-era books. Despite all the restrictions imposed by the Communist regime, the amount of information released on the successful missions should certainly not be underestimated. My philosophy has always been first to analyse what the Russians themselves reported about their missions in TASS announcements, press articles and books – which is often much more than people in the West tend to think. Very helpful in this respect are the digests of TASS communiqués and press articles on space that were published in the Soviet Union on a regular basis [31].

Of course, tracking down both old and new books became vastly easier with the advent of the Internet. I regularly learned of new books from the website of Eastview Publications, an American company specialising in the distribution of Russian books in the West. Although ordering through them is usually costly, it is often the quickest way of obtaining a book, especially if there is a small print run. Nowadays electronic versions of many Soviet-era and Russian-era space books can be downloaded from the Internet, with the most extensive online space library being the one maintained by Sergei Khlynin [32].

A good filing system

Collecting the information is one matter, retrieving it when you require it is another. Therefore I began to keep a database of articles on Soviet space history in journals such as *Spaceflight*, *JBIS*, *Quest*, *Novosti kosmonavtiki* and many others. Although this was a time-consuming job, it has greatly facilitated my search for information, allowing me to find articles in a matter of minutes.

Russian books, magazine articles and the numerous sources now available on the Internet have formed the basis for the space history research that I have been doing for the past twenty years or so. Various personal factors have made it impossible for me to do archival research in Moscow or interview people, so in essence my work has been limited to collecting and analysing the material already published in the Soviet Union and Russia. Therefore much of the credit for the actual research goes to the Russian sleuths. I began publishing the results of my work by occasionally sending letters to the 'Correspondence' sections of *Spaceflight* and *JBIS* and subsequently by publishing articles in these and other magazines and by giving papers at the annual Soviet Forum of the British Interplanetary Society.

GENERAL KAMANIN'S DIARIES

One of my first big projects was to analyse the diaries of General Nikolai Kamanin, which provided an unprecedented glimpse behind the scenes of the Soviet space programme in the 1960s and early 1970s. Clearly, Kamanin kept his journal not so much for personal reflection but to furnish future generations with a true account of events after the veil of secrecy was lifted. Despite being a die-hard Stalinist always loyal to nation and party, Kamanin regularly criticised the lack of openness in media coverage and probably hoped that his diaries would one day rectify this.

The press began to publish excerpts from Kamanin's diaries in the late 1980s, but starting in the mid-1990s the complete set was published in book form through the efforts of his son Lev Kamanin and Sergei Shamsutdinov of *Novosti kosmonavtiki*. There were four volumes covering the periods 1960-1963, 1964-1966, 1967-1968 and 1969-1971 published over a six-year period [33]. I wrote articles about them which appeared in *JBIS* between 1997 and 2002, and together formed a mini-history of the Soviet manned space programme from 1960 to 1971 [34].

A unique viewpoint

The advantage of these diaries was that they were a contemporary record of events, unmarred by the type of errors that typically creep into recollections of events many years or decades ago. Undoubtedly, some editing of the diaries took place, which is evident from the fact that there are differences between the excerpts published in the press and in the books. However, as far as I can tell, these changes were largely of a cosmetic nature and were not intended to rewrite history.

On the other hand, much of what Kamanin wrote must be treated with a degree of caution. His comments on people and events bore a strong personal bias. He did not always have an eye for technical accuracy, and a lot of information was necessarily preliminary, based upon the knowledge that he had at the time of writing. Moreover, Kamanin was not involved in actual space policy decisions, and did not take part in many key meetings which shaped the course of the Soviet space programme. Hence, writing the articles was not just a matter of picking out interesting quotes but also of consulting other sources to verify some of his claims and to fill in the gaps. Because there were virtually no footnotes or editorial notes in the four volumes, a significant amount of background research was needed to better understand many of Kamanin's points. It took me several weeks to go through each of the volumes, taking notes and ordering them by topic, and then several further weeks to write the articles.

The Kamanin diaries marked a milestone in our understanding of the Soviet space programme for several reasons. Firstly, they provided a unique insight into cosmonaut selection, crew assignments and mission planning, and they revealed many mission events that had been completely unknown. Secondly, they shattered the image of the programme having been a centrally managed effort with a transparent hierarchy and structured decision-making policy. This was often illustrated by Kamanin's descriptions of the interdepartmental squabbling between the Air Force and the Strategic Rocket Forces on the one hand and the Air Force and the various design bureaus on the other hand.

Last but not least, the diaries offered a portrait of many key personalities, albeit through the coloured glasses of Kamanin. For instance, while the Soviet propaganda machine depicted the first cosmonauts as impeccable communist heroes, the diaries gave a refreshingly different picture of them as flesh-and-blood people. Of particular interest were some of Kamanin's stories about the personal excesses of cosmonauts

8.9: Cosmonaut boss General Kamanin.

like Yuri Gagarin and Gherman Titov, some of which were reminiscent of those of the Mercury astronauts in Tom Wolfe's *The Right Stuff*.

Stories that come to mind are how Gagarin ended up having a scar above his left eye (the result of a romantic escapade during a vacation on the Black Sea) and the numerous accidents by Titov whilst driving under the influence of alcohol – one of which cost the life of a hitchhiker he had picked up. Also noteworthy are Kamanin's accounts of some of his personal conflicts with key space programme managers and chief designers, particularly Vasily Mishin, whom Kamanin regularly criticised for his drinking habits. It is perhaps surprising that these facts, potentially harmful to the reputation of people who were still alive when the diaries came out, were not edited out. This was an indication that any editing that did take place was superficial. I saw no reason to omit such facts from the summaries because I felt they couldn't do more damage to those personalities than they might already have suffered in Russia.

"SECOND-GENERATION" SOVIET SPACE HISTORY

Kamanin's diaries were part of a "second generation" of publications on Soviet space history that followed the heavily censored "first-generation" accounts of the Communist era. The bulk of them were memoirs by space veterans and cosmonauts. The most significant were the memoirs of Korolev bureau veteran Boris Chertok. These were published in four volumes between 1995 and 1999, and cover virtually all projects of the Korolev design bureau through to the mid-1970s, revealing a vast amount of new facts and figures [35]. Unlike many other memoirs, Chertok's work is clearly based to a large extent on archival material and personal notes, and is written in a fairly objective manner. Of course, these memoirs became available to a wider readership when they were translated into English as part of the NASA History Series under the leadership of Asif Siddiqi. Incorporated into these translations were many changes, additions and corrections made by Chertok on the basis of comments he had received after the original Russian-language series. Sadly, Chertok died in December 2011 at the age of 99, just a few weeks before the final English volume was issued.

Other notable memoirs are those published by Energiya-Buran chief designer Boris Gubanov, TsNIIMash director Yuri Mozzhorin, Korolev bureau veterans Vyacheslav Filin, Konstantin Feoktistov, Oleg Ivanovskiy, Sergei Kryukov and Vladimir Syromyatnikov, and Chelomey bureau veteran Vladimir Polyachenko. There are also numerous memoirs written by lesser luminaries.

Corporate histories

And there were institutional histories. Written by veterans and historians of design bureaus and other space-related organisations, these are inevitably biased. However, they make a significant contribution to our understanding of the Soviet and Russian space programme. Because they are usually not for sale in bookshops, they appear to have been written primarily for people working for the bureaus themselves. But with some effort they can usually be obtained by outsiders. Important examples are those by RKK Energiya, NPO Energomash (the former Glushko bureau), KB Yuzhnoye (the Yangel bureau), NPO Mashinostroyeniya (the Chelomei bureau) and the design bureaus of Myasishchev. The best known of these are the histories of RKK Energiya, three of which have been published so far covering the years 1946-1996, 1996-2001 and 2001-2011 [36]. The first one was a veritable treasure trove of new information on Korolev bureau projects such as Soyuz, the lunar programme, Salyut, Energiya-Buran, etc. A three-volume history of the Military Space Forces was the first to reveal details of many military satellite programmes from the 1960s to the 1980s, and this gave us a better understanding of some Soviet military satellites than their American equivalents [37].

Some of these institutional histories have also disclosed many new facts on post-Soviet era projects. It would be naïve to assume that we know everything about the Russian space projects of the past two decades. Many facts are not reported as they happen, either because they are deliberately kept secret or they are not picked up by the Russian aerospace press. For instance, the last two RKK Energiya histories provide many details about the Russian segment of the International Space Station, the latest Soyuz modifications and the planned successor to Soyuz that weren't even reported by journals such as *Novosti kosmonavtiki*. These days, Russia's military satellites are masked by a veil of secrecy but occasionally something leaks. For instance, a recently published history of the Moscow-based TsNIRTI organisation gave a fairly detailed description of the latest electronic intelligence satellites (Liana) [38]. Maybe the book was never intended for public consumption, but an electronic version showed up on the Internet.

Books by independent researchers

A third category are works written by independent researchers and journalists. One of the earliest cases was a book by Maxim Tarasenko in which he combined his own analytical skills and the available Western literature to produce the first-ever history of Soviet military satellites to be written in the country itself [39]. Were it not for his untimely death in 1999, he would undoubtedly have made many more contributions to the field. Some authors studied the lives and works of famous chief designers such as Sergei Korolev, Valentin Glushko and Mikhail Yangel. Members of the *Novosti kosmonavtiki* team collaborated with other researchers to write two key books. One had detailed biographies of all Soviet and Russian cosmonauts. The other provided a comprehensive history of worldwide piloted space projects, and as such is the most up-to-date book on this topic published anywhere in the world [40]. Another ground-breaking work by Vadim Lukashevich and Igor Afanasyev centred on early Soviet and American cruise missile and spaceplane projects, and had an exhaustive history of the air-launched Spiral spaceplane [41].

Western researchers still active

Regrettably, there has been a dearth of publications written by Russian independent analysts, and their Western counterparts have probably been more active in writing books, basically using the flood of new Russian information to rewrite the Western histories issued during the Soviet era. The first to do so was French space journalist Christian Lardier in his 1992 book *L'astronautique soviétique* [42]. Several years later, Asif Siddiqi synthesised all of this old and new information in his voluminous *Challenge to Apollo*, which also received wide acclaim in Russia itself [43]. A few have also done research on the spot, such as James Harford, who interviewed many space veterans for his Korolev biography [44]. Numerous books on Soviet-Russian space history have come out in the Springer-Praxis series by authors such as Brian Harvey, David Shayler, Rex Hall, Bert Vis, and Grujica Ivanovich. My contribution (a joint effort with Bert Vis) was a book on the history of Energiya-Buran [45].

Finally, there is an increasing body of literature on space history that is aimed at a much broader audience and essentially rehashes well-known facts and stories. In particular, the 50th anniversary of Gagarin's mission inspired many authors to write such books. Unfortunately they are not always technically and historically accurate, and in some cases they bordered on sensationalism.

Of course, much of the "second-generation" history is also being issued as magazine articles and conference papers, although the latter are usually only available as abstracts. Some space history researchers are also active on the Internet. For instance, Vadim Lukashevich has an excellent website on the history of Buran, and Anatoli Zak, a Russian researcher who emigrated to the US in 1993, runs an English-language website that is widely considered to be the best on current and former Russian space projects [46].

"THIRD-GENERATION" SOVIET SPACE HISTORY

Many of the books mentioned above referenced or published excerpts from primary documents such as government and Communist Party decrees, letters and memos written by chief designers, but what was lacking for a long time was access to the primary sources themselves.

A lot of that material (mainly from the 1950s and 1960s) has subsequently been declassified for anyone to see – enabling historians to move on to another phase of Soviet space history research based on documentary evidence. With some effort, it is possible to get access to some of the state archives, although one must have the time and the resources to do so. One of the few Westerners who has done research in the Russian archives and published the results is Asif Siddiqi. He was the first to see a number of key government decrees, such as the critical 3 August 1964 decree that sanctioned the manned lunar effort [47]. Regrettably, design bureau archives remain largely off-limits even to Russian researchers.

Fortunately, some of the declassified primary source material has been published in books. Actually, some of it was released during the Soviet era, the most famous example being *The Creative Legacy of S. P. Korolev* [48]. Published in 1980, it is a collection of documents written by Sergei Korolev and compiled by NPO Energiya historian Georgiy Vetrov. Although clearly edited, the book makes some remarkable revelations, the most conspicuous ones pertaining to early plans for assembly in low Earth orbit using the "Vostok-Zh" and the so-called "Soyuz Complex". For instance, it turned out that the Soyuz was originally to have flown in conjunction with a propulsion unit and several tanker spacecraft, all launched separately. The purpose of the Soyuz Complex was not mentioned, but some hints (notably the phrase that the heat shield had been designed to survive a re-entry at "second cosmic velocity") made it clear that this was an early proposal for a Soviet piloted circumlunar mission. Among the information omitted were the design bureau designators of these spacecraft, which were replaced by fictional names (such as Soyuz-A, Soyuz-B and Soyuz-V for the elements of the Soyuz Complex, which later turned out to be designated 7K, 9K and 11K). It took several months for Westerners to discover the existence of this book, and an article in *JBIS* in 1982 was the first to disclose the assembly plans [49].

Many documents that could not be included in the 1980 book were published in a follow-up volume in 1998 [50]. Also compiled by Vetrov, this contained an amazing collection of documents written by Korolev between 1924 and 1966. A considerable portion of the documents covered the early history of the N-1 rocket, highlighting the debates concerning the type of propellants to be used. What also set this book apart from its predecessor was that many of the documents were accompanied by extensive background commentary.

In 2008 NPO Energomash, Russia's leading engine design bureau, published the declassified correspondence of Valentin Glushko giving new insight into his engine designing work from the mid-1940s to the early 1970s, and illuminating his tenure as chief designer of NPO Energiya between 1974 and 1988 [51].

The publication of government documents on space began in 2008 with a book compiled by former cosmonaut Yuri Baturin that included the texts of numerous key government and Communist Party decrees about the missile and space programmes between 1946 and 1964 [52]. The following year saw the publication of a book with primary documents about the Soviet Union's missile programmes in the 1940s and 1950s [53]. More such publications came out on the occasion of the 50th anniversary of Yuri Gagarin's mission in 2011, including two books featuring documents about the early days of the Vostok programme from government and design bureau archives [54]. One other book has a wonderful collection of government documents between 1956 and 1965 [55]. It includes not only the cover letters of the decrees, but also the crucial supplements to the decrees, which are often more informative than the cover letters themselves.

CONCLUSIONS

Some twenty-five years after the introduction of *glasnost*, our understanding of the history of the Soviet and Russian space programme has vastly improved. Some may think that the golden age of sleuthing is over and that little work remains to be done in the field. Whilst it is probably true that all the 'big' revelations have been made and we are unlikely to discover any more completely unknown projects, many key questions of a historical nature remain in regard to both the manned and unmanned programmes. The increasing availability of primary sources is making it possible to enter a new stage of research, but as yet only a small fraction of these sources have been explored.

Therefore, one can only hope that more and more (especially Russian) researchers will delve into the archives, talk to surviving veterans and publish the results of their research. Important information about the history of the Soviet space programme can also be gleaned from US reconnaissance satellite pictures of Soviet launch sites and from declassified CIA documents. In contrast to what some may think, the sleuthing business is not in its death throes, and both Western and Russian sleuths have their work cut out for them for many more years to come. Perhaps we should say that the golden age of sleuthing is only just beginning.

References

1. The revelations about the Gagarin team were made in a series of articles in *Izvestia* in 1986, which later that same year was reprinted in a booklet: Ya. Golovanov, *Kosmonavt N1*, Moscow: *Izvestiya*, 1986; "Female Selection of 1962" (in Russian), *Rabotnitsa*, 10/1987.
2. M. Pastukhova, "Brighter Than Any Legend" (in Russian), *Ogonyok*, December 1987 (nr. 49); A. Bolotin, "Site nr. 10" (in Russian), *Ogonyok*, April 1989 (nr. 16).
3. G.V. Petrovich (ed.), *Kosmonavtika. Malenkaya entsiklopediya*, Moscow: *Sovetskaya Entsiklopediya*, 1969. Translated into English the same year by Mir Publishers as *The Soviet Encyclopedia of Space Flight*. G.V. Petrovich was the pseudonym of Valentin Glushko. The second, updated edition appeared in 1970 under Glushko's real name.
4. A. Volskiy (ed.), *Kosmodrom*, Moscow: Voyennoye izdatelstvo Ministerstva oborony SSSR, 1977, p. 126.
5. L. Kamanin, "From the Earth to the Moon and Back" (in Russian), *Poisk*, July 1989 (nr. 12).
6. G. Reznichenko, *Kosmonavt-5*, Moscow: Politizdat, 1989, p. 98.
7. S. Leskov, "How We Didn't Fly to the Moon" (in Russian), *Izvestiya*, 18 August 1989.
8. S. Leskov, *Kak my ne sletali na lunu*, Moscow: Panorama, 1991.
9. N. Kamanin, "I Would Never Have Believed Anyone" (in Russian), *Sovetskaya Rossiya*, 11 October 1989.
10. A. Tarasov, "Flights in Dreams and Reality" (in Russian), *Pravda*, 20 October 1989.
11. See for instance: J.N. Wilford, "Moon Race Was Real, Moscow Now Admits", *International Herald Tribune*, 19 December 1989; C. Covault, "Soviet Manned Lunar Mission Plan Used Modified Soyuz Spacecraft", *Aviation Week & Space Technology*, 8 January 1990, p. 44.
12. M. Rebrov, "A Version of the 'L' Project" (in Russian), *Krasnaya zvezda*, 30 October 1993.
13. J. Parfitt, A. Bond, "The Soviet Manned Lunar Landing Programme", *JBIS*, vol. 40 (1987), pp. 231-234.
14. S. Young, "Soviet Union Was Far Behind in 1960's Moon Race", *Spaceflight*, January 1990, pp. 2-3.
15. M. Rebrov, "That Is How Things Were. The Difficult Fate of the N-1 Project" (in Russian), *Krasnaya zvezda*, 13 January 1990.
16. "Commentaries on N-1 Booster and Manned Lunar Mission Programme", *JPRS Report/ USSR: Space*, 15 May 1990, pp. 45-51.
17. I. Kuznetsov, "The Flight That Didn't Take Place" (in Russian), *Aviatsiya i kosmonavtika*, 8/1990, pp. 44-45.
18. V. Mishin, "Why We Didn't Fly to the Moon" (in Russian), *Kosmonavtika, astronomiya*, 12/1990.
19. C. Covault, "Soviet Union Reveals Moon Rocket Design that Failed to Beat

U.S. to Lunar Landing", *Aviation Week & Space Technology*, 18 February 1991, pp. 58-59; "Soviet Moon Rocket Revealed", *Spaceflight*, March 1991, p. 78.

20 "The Moon Programme That Faltered: Vasily Mishin Outlines Soviet Manned Lunar Project: N-1/L-3", *Spaceflight*, January 1991, pp. 3-4.

21 "Mishin Monograph on Failure of Soviet Manned Lunar Program", *JPRS Report/USSR: Space*, 12 November 1991. A colourfully illustrated French translation appeared two years later: *Pourquoi nous ne sommes pas allés sur la lune*, Toulouse: Cépaduès-Editions, 1993.

22 The digest with the smallest circulation was known as *Osobye zakrytye pis'ma* ("Special Secret Letters"). There were several other, more widely disseminated TASS digests of the foreign press, including one informally known as "White TASS" (Belyy TASS).

23 V. Molchanov, "About Those Who Did Not Go Into Orbit" (in Russian), *Kosmonavtika, astronomiya*, 10/1990.

24 I. Afanasyev, "Unknown ships" (in Russian), *Kosmonavtika, astronomiya*, 12/1991.

25 "Development of Soviet spacecraft for Manned Missions", *JPRS Report/Central Eurasia: Space*, 27 May 1992.

26 Yu. Biryukov, "The Long Road To A Space Journal" (in Russian), *Novosti kosmonavtiki*, 9/2001, p. 6.

27 V. Agapov, "Anniversary of the launch of the first artificial Earth satellite in the DS series" (in Russian), *Novosti kosmonavtiki*, 6/1997, pp. 54-64; V. Sorokin, "Amber story" (in Russian), *Novosti kosmonavtiki*, 17/1997, pp. 57-64, 18-19/1997, pp. 91-99.

28 K. Lantratov, "The Star of Dmitry Kozlov" (in Russian), *Novosti kosmonavtiki*, 3/1997, pp. 50-55, 4/1997, pp. 82-84, 5/1997, pp. 81-86, 6/1997, pp. 74-80. Translated into English as "Soyuz-Based Manned Reconnaissance Spacecraft" in *Quest*, vol. 6 nr. 1 (spring 1998), pp. 5-21.

29 Twelve articles on the history of the R-7 and derived versions appeared in *Spaceflight* under the title "Soviet Rocketry That Conquered Space" between August 1995 and April 2001.

30 Yu. Markov, *Kurs na Mars*, Moscow: Mashinostroyeniye, 1989.

31 This series was called *Osvoyeniye kosmicheskogo prostranstva v SSSR*, translated by NASA as *The Conquest of Outer Space in the USSR*. The first two volumes covered the period 1957-1967 and 1967-1970 and from 1971 on the digests appeared on an annual basis.

32 http://epizodsspace.no-ip.org/

33 N. Kamanin, *Skrytyy kosmos. Kniga pervaya. 1960-1963*. Moscow: Infortekst, 1995; N. Kamanin, *Skrytyy kosmos. Kniga vtoraya. 1964-1966*. Moscow: Infortekst, 1997; N. Kamanin, *Skrytyy kosmos. Kniga tretya. 1967-1968*. Moscow: Novosti kosmonavtiki, 1999; N. Kamanin, *Skrytyy kosmos. Kniga chetvyortaya. 1969-1971*. Moscow: Novosti kosmonavtiki, 2001.

34 B. Hendrickx, "The Kamanin Diaries 1960-1963", *JBIS*, vol. 50 (1997), pp. 33-40; "The Kamanin Diaries 1964-1966", *JBIS*, vol. 51 (1998), pp. 413-440; "The

Kamanin Diaries 1967-1968", *JBIS*, vol. 53 (2000), pp.384-428; "The Kamanin Diaries 1969-1971", vol. 55 (2002), pp. 312-360.

35   B. Chertok, *Rakety i lyudi*, Moscow: Mashinostroyeniye, 1995; B. Chertok, *Rakety i lyudi. Fili-Podlipki-Tyuratam*, Moscow: Mashinostroyeniye, 1996; B. Chertok, *Rakety i lyudi. Goryachiye dni kholodnoi voiny*, Moscow: Mashinostroyeniye, 1997; B. Chertok, *Rakety i lyudi. Lunnaya gonka*, Moscow: Mashinostroyeniye, 1999.

36   Yu. Semyonov (ed.), *Raketno-kosmicheskaya korporatsiya Energiya imeni S.P. Korolyova 1946-1996*, Moscow: RKK Energiya, 1996; Yu. Semyonov (ed.), *Raketno-kosmicheskaya korporatsiya Energiya imeni S.P. Korolyova nu rubezhe dvukh vekov 1996-2001*, Moscow: RKK Energiya, 2001; V. Lopota (ed.), *Raketno-kosmicheskaya korporatsiya Energiya imeni S.P. Korolyova v pervom desyatiletii XXI veka*, Moscow: RKK Energiya, 2011.

37   V. Favorskiy, I. Meshcheryakov, *Voyenno-kosmicheskiye sily. Kniga 1: kosmonavtika i vooruzhonnye sily*, Moscow, 1997; V. Favorskiy, I. Meshcheryakov, *Voyenno-kosmicheskiye sily. Kniga 2: stanovleniye Voenno-kosmicheskikh sil*, Moscow: 1998; V. Ivanov (ed.), *Voyenno-kosmicheskiye sily. Kniga 3: Voyenno-kosmicheskiye sily v period radikalnoi perestroiki kosmicheskoi deyatelnosti Rossii*, Moscow, 2001.

38   B. Lobanov (ed.), *TsNIRTI 65 let*, Moscow, 2008.

39   M. Tarasenko, *Voyennye aspekty sovetskoi kosmonavtiki*, Moscow: Nikol, 1992.

40   I. Marinin, S. Shamsutdinov, A. Glushko, *Sovetskiye i rossiyskiye kosmonavty. 1960-2000.* Moscow: Novosti kosmonavtiki, 2001; Yu. Baturin (ed.), *Mirovaya pilotiruemaya kosmonavtika. Istoriya. Tekhnika. Lyudi.* Moscow: RTSoft, 2005.

41   V. Lukashevich, I. Afanasyev, *Kosmicheskiye krylya*, Moscow: OOO LenTa Stranstviy, 2009.

42   C. Lardier, *L'astronautique soviétique*, Paris: Armand Colin, 1992.

43   A. Siddiqi, *Challenge to Apollo: The Soviet Union and the Space Race 1945-1974*, Washington D.C: NASA, 2000.

44   J. Harford, *Korolev*, New York: John Wiley, 1997.

45   B. Hendrickx, B. Vis, *Energiya-Buran: The Soviet Space Shuttle*, Chichester: Praxis Publishing, 2007.

46   www.buran.ru; www.russianspaceweb.com

47   The results of his research into the origins of the Soviet space programme were published in: A. Siddiqi, *The Red Rockets' Glare: Spaceflight and the Soviet Imagination, 1857-1957*, New York: Cambridge University Press, 2010; For an article on the 3 August 1964 decree, see: A. Siddiqi, "A Secret Uncovered: The Soviet Decision To Land Cosmonauts on the Moon", *Spaceflight*, May 2004, pp. 205-213.

48   M. Keldysh, *Tvorcheskoye naslediye akedemika Sergeya Pavlovicha Korolyova. Izbrannye trudy i dokumenty.* Moscow: Nauka, 1980.

49   C. Wachtel, "Design Studies of the Vostok-J and Soyuz Spacecraft", *JBIS*, vol. 35 (1982), pp. 92-94.

50   G. Vetrov (ed.), *S.P. Korolyov i ego delo. Svet i teni v istorii kosmonavtiki*, Moscow: Nauka, 1998.

51   *Izbrannye raboty akademika V.P. Glushko* (3 volumes), Khimki: 2008.
52   Yu. Baturin (ed.), *Sovetskaya kosmicheskaya initsiativa v gosudarstvennykh dokumentakh 1946-1996 gg.,* Moscow: RTSoft, 2008.
53   V. Ivkin, G. Sukhina (ed.), *Zadacha osoboi gosudarstvennoi vazhnosti. Iz istorii sozdaniya raketno-yadernogo oruzhiya i Raketnykh voisk strategicheskogo naznacheniya (1945-1959 gg.)*
54   V. Davydov (ed.), *Pervyy pilotiruemyy polyot* (two volumes), Moscow: Rodina Media, 2011; L. Uspenskaya et. al., *Chelovek. Korabl. Kosmos.*, Moscow: Novyy khronograph, 2011.
55   S. Kudryashov (ed.), *Sovetskiy kosmos,* Moscow: Vestnik Arkhiv Prezidenta Rossiyskoi Federatsii, 2011.

9

People and archives

By Asif A. Siddiqi

What I'd like to do here is to tell three stories that I hope are loosely connected. The first is a brief personal account of how I got interested in the history of the Soviet space programme, joining others in the West who were trying to uncover its secrets. When I began to study the Soviet space programme in the late 1970s, the names of its major architects were little known. In the second part of the essay, I explain how the names and identities of the most prominent designers behind the programme came to public attention. They include Sergei Korolev, Valentin Glushko, Mikhail Yangel, Vladimir Chelomey, and Vasily Mishin. Whilst I was not personally involved in this sleuthing – which occurred mostly in the 1960s and 1970s – the process of pulling back the curtains was very influential in my own work. Finally, in the concluding section, I build on the first two sections – the personal and the investigative aspects of sleuthing – and present some reflections on my journey into the *archives* in the post-Soviet period. I show how some of my work has helped to deepen our knowledge of the lives and works of men like Korolev, Glushko and Chelomey, and how my own voyage into the depths of the programme has come full circle: I have now met many veterans who worked with men like Korolev and Chelomey, giants whose very lives and works I was trying to uncover.

A PERSONAL JOURNEY

Like many others who were drawn to the study of the Soviet space programme, I was captivated at an early age. My first memory of a Soviet space "event" was in 1977 when I was aged eleven and living in Manchester, England. I have a distinct memory of cutting out a newspaper story on Soyuz 25, which had failed to dock with the new Salyut 6 space station. Later, I watched in wonder as British television showed grainy (but colour!) footage of cosmonauts Romanenko and Grechko inside Salyut 6. My family soon returned to Bangladesh for several years, but my interest only grew in leaps and bounds. To the alarm of my parents, I obsessively listened to

9.1: An introduction to Soviet spaceflight with *Transfer in Orbit* (left), the *Observer's Book of Manned Spaceflight* and the *Soviet Encyclopedia of Spaceflight*.

the English-language radio broadcasts of Radio Moscow. Inevitably, every single day there would be a news report on what the cosmonauts on Salyut 6 were up to. It was an exciting time for me and I was a fast learner. Fortunately, in Dhaka we lived very close to the Russian Cultural Centre, which had a superb library where I first perused through many English-language books on the Soviet space programme, including one of my all-time favourites, *Transfer in Orbit*, an illustrated book on the Soyuz 4 and 5 docking in January 1969 [1].

In 1978 my father bought me a copy of the *Soviet Encyclopedia of Spaceflight*, another English-language book edited by one "G. V. Petrovich" which I prized [2]. This lovely volume, originally published in mid-1969, was obviously intended for the foreign market. While the illustrations were of poor quality, this was compensated by the essay entries, especially on personalities from the history of Soviet cosmonautics. And this "Petrovich" fellow had also included a complete list of all Soviet launches from 1957 to 1969 – an absolute treasure for me as I pored over the various satellites, mysteriously all named "Kosmos".

Following Salyut missions

The late 1970s was a time of rebirth for the Soviet space programme: each long-stay mission to Salyut 6 set new endurance records; the station itself could receive two spaceships simultaneously, and to top it all off, there was a new cargo version of the Soyuz named Progress that could remotely dock with the station. At 3:00 p.m. every day I wrote copious notes from Radio Moscow about magnificent cosmonauts with enigmatic names such as Romanenko, Grechko, Kovalenok, Ivanchenkov, Lyakhov, Ryumin, Popov and so on.

But even at this early stage, I sensed that the official Soviet pronouncements told only part of the whole story. I had a treasured copy of *The Observer's Book of Manned Spaceflight* by Reginald Turnill, the very first space book I ever purchased in Manchester [3]. More than anything, this wonderful little book hinted at obfuscations and occlusions in the received story of the Soviet space programme. For example, why weren't the names of cosmonauts given in advance? And who were the backup crews? My interests at that time gravitated less towards the hardware than the people behind this enigmatic adventure, and in particular the cosmonauts. I started to piece together possible backup crews for Soviet space missions, but the absence of Western literature in Bangladesh made this very difficult. All I had were Soviet journals such as *Ogonek*, *Soviet Life*, and such, some of which were actually published in Bengali.

My early attempts at 'space sleuthing'

In 1981 my parents bought me a copy of *Red Star in Orbit* by James E. Oberg [4]. It would be an understatement to say that my world view was transformed. I practically memorised the entire book, and re-read a hundred times the end section where Oberg listed still unanswered questions. Oberg's book also introduced as a real character to my worldview, the person of Sergei Korolev, who had been prominent but not unduly so in "Petrovich's" encyclopedia. Further acquisitions followed, most notably the Congressional Research Service's series of volumes under the general title *Soviet Space Programs* [5]. In leaps and bounds, I internalised the minutiae of the Soviet space programme, to the alarm of my family who must have thought I'd gone insane.

By 1982, I wrote a history of the Soviet space programme running to almost fifty pages (which I still have) based on various English-language sources. Only much later did I realise that my work was still far, far behind such Western luminaries as Geoffrey E. Perry (1927-2000), Charles S. Sheldon (1917-1981), James E. Oberg, Nicholas L. Johnson, and Phillip S. Clark. But this early attention to detail, repeating (albeit rather poorly) what others had done before, was immensely useful in laying a foundation. Once I permanently moved to the United States in 1985, aged eighteen, I was able to make use of a vast canon of English-language literature on the Soviet space programme and start to draw my own conclusions. As is no doubt evident from this volume, much of this ground-breaking work that I made use of was published in the pages of *Spaceflight* and the *Journal of the British Interplanetary Society*, two journals which were the gospel to me in the 1980s. Exposure to the research of the British contingent – especially Rex Hall (1946-2010), Phillip Clark, R. F. Gibbons, Gordon R. Hooper, and Neville Kidger (1953-2009) – was crucial in filling in the gaps.

Glasnost transforms everything

If the first transformative event in terms of my Soviet space "education" was my initial exposure to the Soviet space programme in the late 1970s, something equally transformative occurred in the late 1980s – *glasnost*. This period of "openness" let

9.2: With veteran cosmonaut Vladimir Shatalov, one of the leading candidates for a Soviet moonlanding, at Star City in May 2006.

loose the floodgates of information on the history and present of the Soviet space programme. It was hard to keep up with the deluge, particularly in the pages of not only *Spaceflight* and *JBIS* but also other short-lived publications such as *Zenit* and *Spaceflight News*. What grabbed most of my attention at the time (and probably for others too) were the revelations concerning the Soviet manned lunar programme. Soviet censors first allowed references to the programme in the summer of 1989.

Using some of the initial trickle, in 1991 I wrote and "self-published" a 100-page brochure called *The Soviet Piloted Lunar Programme*, and sent it to a few friends, but the information was coming at such a rate that it was obsolete after a few months. At this time, it was evident that I was learning about the Soviet lunar programme from English-language sources, particularly the work of such pioneering sleuths as Clark (a hero of mine) and Johnson, the latter of whom published *The Soviet Reach for the Moon* in 1994 [6]. But the more I read, the more curious I got. Perhaps I ought to go to the horse's mouth and track down the original Russian-language sources used by others? Using translated Russian sources, in 1994 I published my first articles on the history of the Soviet space programme. One was a two-parter on the organisation of their effort in *Spaceflight*, and the other was a lengthy scoop on the so-called Nedelin disaster in the U.S. magazine *Quest* [7].

Decision to study Russian

My decision in 1991 to use Russian sources instead of English-language ones led me down an unexpected path when I set out to write a comprehensive monograph on the history of the Soviet space programme based largely on Russian-language sources.

It was at this time that I made contact with another Russian space sleuth who had a profound impact on my future trajectory. In early 1993 I began communicating via e-mail with Dennis Newkirk, a young American writer who had recently published the *Almanac of Soviet Manned Space Flight* [8]. Dennis began to send me everything that he was collecting from writers all around the world, and soon we began to collaborate on a history of the Functional Cargo Block (FGB), a spacecraft that was intended to serve as a key element of the new International Space Station (ISS). Dennis and I dug quite deep, collecting an enormous amount of data and determined the intricate history of the FGB and its former role as part of the Transport-Supply Ship (TKS) of the Almaz programme. We also identified the Khrunichev factory as being part of a larger and convoluted network of design bureaus and factories which had been formerly linked to Vladimir Chelomey and Vladimir Myasishchev, two of the most prominent Soviet aerospace chief designers. We presented the results of our research at the annual meeting of the Society for the History of Technology (SHOT) in 1995 [9].

More than anything, Dennis encouraged me to keep working on my book about the history of the Soviet space programme. He put me in touch with Glen Swanson, editor of *Quest*, a fledgling new journal on the history of spaceflight that published many new revelations on the history of the Soviet space programme. Glen in turn helped me contact Roger Launius, the chief historian at NASA who serendipitously found some value in my notion to write a comprehensive history. Roger helped me to bring to fruition five years of dedicated work by having the NASA History Office publish in the year 2000 my *Challenge to Apollo: The Soviet Union and the Space Race, 1945-1974* [10]. Even after publication, I kept in touch with Dennis although owing to his full-time job and family obligations (he married and had a son) he was able to devote less and less time to sleuthing himself. Tragically, on 12 April 2012 Dennis passed away in Barrington, Illinois (a suburb of Chicago) from cancer. He was only 47 years old [11].

KHRUSHCHEV'S SECRECY

One of the biggest challenges in writing *Challenge to Apollo* was to offer a human side to the story of the Soviet space programme. This meant investigating in depth the lives of the principal players. I wanted to go beyond simply regurgitating information about Korolev and Glushko, and dig deeper. When my interest was first awakening in 1977, I had no idea who actually was "behind" it all. Korolev, Glushko and Yangel are now widely known, but until the late 1980s such names had little meaning for most of us. The books I owned (such as Turnill's) all seemed to communicate only a general impression, perhaps mentioning Korolev but only in the vaguest terms. Given the dearth of details about these men, my interest was piqued. And more recently, I have kept returning to a simple question: How did these names come to be known? Clearly they didn't suddenly appear into public view. What was the process? Who were the Westerners who found them out? This then is the subject of the next part of my essay.

Sergei Pavlovich Korolev (1907-1966) is rightly considered the founder of the Soviet space programme. Not surprisingly, there has been an enormous amount of scholarship devoted to his life and activities. As is well-known, during his lifetime the Soviet government went to extraordinary lengths to ensure that his identity (and indeed those of other prominent space designers) remained unknown. In a speech in 1958, Soviet Communist Party First Secretary Nikita Khrushchev famously said:

> The Soviet atomic specialists or the experts who created the intercontinental rocket and the artificial earth satellites have no complaints about the socialist state. . . . The Soviet government rewards them; they are materially well taken care of and many of them have received Lenin Prizes and the Order of Hero of Socialist Labour. They 'suffer' a little only in one respect: for the time being they are anonymous to the outside world. They live under the title: 'Scholars and engineers working on atomic and rocket technology.' But who these people really are is now widely unknown. For those who created the rockets and artificial earth satellites we will raise an obelisk and inscribe their glorious names on it in gold so they will be known to future generations in the centuries to come. Yes, when the time comes photographs and the names of these glorious people will be published and they will become broadly known among the people. We value and respect these people highly and assure their security from enemy agents who might be sent to destroy these outstanding people, our

valuable cadres. But now, in order to guarantee the security of the country and the lives of these scholars, engineers, technicians, and other specialists, we cannot make their names public or print their pictures. [12]

Since that time, it has become an oft-repeated truism in history books that during Korolev's lifetime there was no indication of the true identity of the so-called Chief Designer. Occasionally Westerners might hear about this enigmatic man, but until his death no one knew who he was. This is only partially true. While his identity was a closely guarded secret in the Soviet Union, his name was actually widely known in the West *before* his death. A number of Western analysts, by intelligent deductions, had managed to identify him as the "Chief Designer". In addition, these revelations appeared not only in obscure media but in major publications such as *Spaceflight*, the *New York Times*, and *Fortune* magazine.

The designer who came in from the cold

Korolev did not live an anonymous life. In fact, prior to his career in the rocket and space business, he was a fairly well-known designer of gliders in the 1930s [13]. By the time that he was in his early twenties, newspapers and magazines were already writing about him. He, in turn, authored a number of important articles on aviation. For example, the official military newspaper, *Krasnaya zvezda* (*Red Star*) published several pieces on the impressive performance characteristics of his SK-3 glider, also coincidentally named *Krasnaya zvezda* [14]. In 1931 his SK-4 was featured on the cover of the official journal of the Soviet Air Force, *Vestnik vozdushnogo flota* (*Journal of the Air Fleet*) as a "new Soviet long-range light-aircraft designed by S. Korolev" [15]. In other words, by the early 1930s Korolev's name was well-known amongst a fairly broad group of aviation enthusiasts in the Soviet Union. His first and only monograph, *Raketnyy polet v stratosfere* (*Rocket Flight into the Stratosphere*), published in late 1934, was not a book about space exploration but about high-speed rocket-powered aviation. The editor's introduction in the book stated confidently:

> The author, pilot-engineer S. P. Korolev, in his work depicts the significance of the struggle to achieve great flight altitudes and underscores the capabilities of reactive flying vehicles as the most important means to achieve this goal. In the work, [the author] deals with experiments carried out with reactive flying vehicles; for the first time in our literature the design of modern reaction engines are shown and problems outlined which will allow the accomplishment of reactive flight of humans into the stratosphere. [16]

The book was positively reviewed in a number of different places, bringing his name to a wide audience [17].

As Korolev became deeply involved in military rocketry at the Reactive Scientific-Research Institute (RNII) in the 1930s, his public profile faded. There were few open details of his work on rockets at the time. Most of his research was focused on cruise missiles and rocket-powered gliders, both of which had military applications. The last article under his own name published before the onset of World War II was issued in 1937 when he wrote a brief review of several recent books on stratospheric aviation [18]. His arrest and subsequent incarceration, beginning 1938, ensured that his name was entirely absent from any public discussion during the war [19]. Unless one was (in)famous, convicted prisoners were hardly mentioned in the Soviet press. Korolev simply disappeared, both in body and in mind, and his brief fame in the early 1930s became a forgotten footnote in the history of Soviet aviation.

As is well-known, Korolev was released from prison in 1944 and two years later was appointed one of several Chief Designers at a new rocket development institute (NII-88) located in the northeastern Moscow suburb of Kaliningrad. His position and work were top secret, and like almost all the other designers of weapons systems, he kept out of the public eye. In the decade between his appointment as Chief Designer and the launch of Sputnik his name *did* appear in official print, but only in a context that would not allow anyone to guess his "real" duties. In September 1957, just a few weeks prior to Sputnik, he delivered a prominent speech to commemorate the 100th birthday of the founding theorist of Soviet cosmonautics, Konstantin Tsiolkovsky. An abridged version was published on page 2 of *Pravda* as part of a special tribute to the late Tsiolkovsky. Korolev carefully, and perhaps intentionally, observed that "in the near future, for scientific purposes, the first trial launches of artificial satellites of the Earth will take place in the USSR and the USA". (Both nations had announced their intention to put up a satellite to mark the International Geophysical Year.) He signed simply as "S. Korolev, Corresponding Member of the USSR Academy of Sciences" [20]. In fact, this would be the very last article that Korolev published under his own name during his lifetime.

CIA sources wrong

What did Western observers know about Korolev during this period? One would expect that intelligence analysts at the Central Intelligence Agency would have had the best opportunity to uncover Korolev's identity. For certain, he would have been a high value target, given his prominence as the Chief Designer of the Soviet ICBM. Yet strangely enough, the declassified records of the CIA rarely mention his name, and when his name occurs it is never as a leading scientist or a chief designer. Most of the information on Soviet rocketry came from interviewing German specialists. Many of these men had known Korolev quite well, especially during the time that the Germans helped the Soviets to establish assembly and production of local versions of the V2 missile. By 1953 almost all of them had returned to the German Democratic Republic (East Germany) and subsequently the Federal Republic of Germany (West Germany). But anticipating that these specialists would eventually be deported, after late 1947 the Soviets had isolated them from mainstream rocket development [21].

Once these men were back in the West, the intelligence agencies sought them out and interviewed them for information on Soviet missiles. Some of this information actually leaked out in the 1950s. For example, the wife of Helmut Gröttrup, probably the most prominent German engineer who had worked with the Soviets, published a colourful memoir in German. Filled with inaccuracies and exaggerations, the memoir nevertheless provided a peek into the social history of German engineers kidnapped to work on Soviet rockets [22]. The CIA, which had better information, kept its data under wraps and most of this analysis was unknown until the end of the Cold War, when it declassified thousands of documents. These clearly show that the agency was interested in identifying important personalities. In a classified report issued in 1953, the agency included a list of personalities that included Boris Chertok, Lev Gaydukov and Yuri Pobedonostsev. Yet these names were listed in very general terms and it is evident that the CIA knew very little about what these men were actually doing at the time. "Korolov" (sic), for example, was listed only as "former deputy at Bleicherode" [23].

In another report on Soviet guided missile development that was issued in 1960 (presumably after more interviews with the returned Germans) the CIA once again listed several important leaders: Boris Chertok, Lev Gonor, Boris Konoplev, Vasily Mishin, Yuri Pobedonostsev, Konstantin Rudnev, Mikhail Ryazansky, Mikhail Tikhonravov, Georgi Tyulin and Leonid Voskresensky. Although this would seem to be a very good list of the top designers in the late 1950s, most of the information was dated to the late 1940s and the CIA had no information about what these men were up to in more recent years. About Korolev, the agency wrote: "[i]t was generally agreed that the most talented Soviet engineer-designer at NII-88 was a Colonel Sergei P. Korolev." As with the others, the CIA had no information from the late 1950s. In the 1960 report they noted that Korolev was a "chief designer as of 1951" but could not say what he was doing after that year [24]. And even more remarkably, there was no mention of Valentin Glushko, Chief Designer of the rocket engines that powered the first Soviet ICBM.

West fooled by 'front men'

Overall, while the CIA's understanding of the technological aspects of the Soviet space programme had some basis in reality, its view of the management was poor, especially in the early years. For example, as late as April 1961 the CIA claimed that the Soviet space programme was directed by something known as the "Inter-agency Commission for Interplanetary Communications under the Astronomy Council of the Academy of Sciences" [25]. Such a commission had actually been announced by the Soviet media as long ago as 1954 but it was mentioned less after Sputnik; only after the Cold War did it turn out that this supposed Commission was actually a "cover" institution set up by the Soviets to publicise their participation in the International Geophysical Year [26]. In the United States, the two best-known *public* personalities linked with Soviet spaceflight actually had little to do with either Sputnik or Vostok – Academicians Leonid Sedov and Anatoli Blagonravov were, respectively, prominent scientists in gas dynamics and machine science. Both men travelled abroad frequently and were quoted widely in Western European and American newspapers as scientific and technical leaders of the Soviet space effort. With very little to go on, journalists simply assumed that Sedov and Blagonravov worked at the highest levels. However, these men only had a peripheral relationship with the heart of the space programme. The renowned Soviet journalist Yaroslav

Golovanov later wrote that these "public" spokespersons "were so ensnared by what they had signed about not disclosing governmental secrets, that they uttered only banalities" [27].

Chief Designer's secret identity

After Sputnik, Korolev's name disappeared from public view. Nevertheless, he and Glushko wrote frequently in the official Soviet press as "Professor K. Sergeyev" and "Professor G. V. Petrovich" respectively, simply by playing with their names: Sergei Pavlovich Korolev and Valentin Petrovich Glushko. French space sleuths Christian Lardier and Claude Wachtel pioneered the study of these pseudonyms and identified dozens of fake names used by real designers. The end of the Cold War confirmed many of their guesses, and Lardier was able to present a complete summary of this research in 1996 [28].

Korolev's articles usually appeared in the official Soviet party newspaper *Pravda* every new year's day – a very high honour accorded to influential Soviet dignitaries. His first article using his pseudonym was published in December 1957, only weeks after the spectacular successes of the first two Sputniks. Note that he had written an article under his *real* name just two months earlier to mark the 100th anniversary of Tsiolkovsky's birth; he had returned to the "black world". Although Korolev's many pseudonymous articles were very general in nature, they typically anticipated new Soviet developments in space exploration the following year; in other words, they usually dealt with the vast possibilities of the future [29]. Glushko's articles, on the other hand, were often historical in nature, and focused on the activities of the Gas Dynamics Laboratory (GDL) in the late 1920s and early 1930s where he served his apprenticeship. For example, on the occasion of the joint flight of Andrian Nikolayev on Vostok 3 and Pavel Popovich on Vostok 4 in August 1962, Glushko wrote a long two-part article in the newspaper *Komsomol'skaya pravda* as "G. Petrovich" on the early history of Soviet rocketry [30]. Glushko also wrote many articles for the official journal of the Academy of Sciences which would be closely scrutinised by Western analysts for information on Soviet space technology such as the early Sputniks or the Proton booster.

Korolev named in the West

Yet, even during this period, there were already rumours about Korolev and Glushko in the West. Korolev's name (and his possible role in the Soviet space programme) first appeared in print in the Western press through leaks from Soviet defectors or from American journalists stationed in Moscow. In September 1961, a former Soviet citizen, Grigory Aleksandrovich Tokaty-Tokaev (1910-2003), gave a talk on Soviet spaceflight at the British Interplanetary Society (BIS) in London. Tokaty had been a representative of the Soviet Air Force in occupied Germany after the war in 1945. Although he didn't have much contact with the Nordhausen Institute responsible for recovering German rocket technology, he picked up a particularly unique assignment. In 1947, Stalin assigned him to lead a small team to kidnap the Austrian aeronautics pioneer Eugen Sänger and bring him back to the Soviet Union. The idea was to have Sänger work in a Soviet design bureau (OKB-3 under Gherman Moishev) and help the Soviets to develop the so-called antipodal bomber for intercontinental flight [31]. While on his mission in occupied Germany in late 1948, Tokaty defected to Britain where he lived for the rest of his life [32].

Tokaty had had little direct contact with the Soviet missile programme, and later grossly exaggerated his role in the postwar missile effort by claiming that he was the "chief rocket scientist" of the Soviet Union, when in fact there was no such position. Some of his information was also clearly wrong or exaggerated [33]. Yet Tokaty did know several key facts about the Soviet missile programme which were unknown to the general public. He was the first person in the West to openly suggest that Sergei Korolev was involved with the Sputniks and Vostok. In his 1961 lecture to the BIS, Tokaty said Korolev was "*one of the* chief designers of rockets for carrying Sputniks and Vostok capsules". He also mentioned Valentin Glushko, but was unsure of his exact role in the successes of Sputnik and Vostok. The text of Tokaty's speech was published in several different places but few people paid attention [34].

Reports on Korolev and Glushko's true identities continued to emerge from time to time in the early 1960s. For example, in November 1963, during the wedding of cosmonauts Andrian Nikolayev and Valentina Tereshkova, Western correspondents were invited to the reception – and learned through informal conversation that two important scientists from the Soviet space programme were in attendance, "S. P. Korolev" and "V. P. Glushko". Shortly thereafter, Theodore Shabad (1922-1987), an enterprising journalist for the *New York Times*, published a story identifying Korolev and Glushko as "likely two figures in the Soviet space programme". He was not sure which one of the pair was the "Chief Designer of Rocket-Space Systems" and which was the "Chief Designer of Rocket Engines", but it appeared that they were of equal importance. Shabad incorrectly claimed that Glushko had worked with Soviet rocket engineer Fridrikh Tsander in the 1930s [35].

Conclusive identification

Around this time, quite independently, the Aerospace Information Division (AID) at the Library of Congress came to the same conclusion concerning Korolev's identity. Where the *New York Times* had felt unsure of their guess, AID was the very *first* Western organisation to confidently pinpoint the identity of the mysterious "Chief Designer". And it bears repeating that they identified Korolev long before the CIA. The Library of Congress based its research on a detailed analysis of all of the open Russian-language literature on rocketry between 1934 and 1964 [36].

The strategy they used to identify Korolev was rather interesting. In 1962, the publisher "Sovetskaya rossiya" issued a book by the title *Nashi kosmicheskiye puti* (*Our Paths in Space*) containing various essays and documents from the early years of the Soviet space programme. It matched the usual archetype of early Soviet space publications, with pages and pages of press releases, descriptive passages, laudatory poems in honour of the socialist cause, and few if any details of actual space flights. But a careful reading of the articles showed that one must be creative in seeking the secrets of the Soviet space programme. One of the articles in the book, 'Vse li my znayem o tsiolkovsom?' ('Do we know everything about Tsiolkovsky?') was by Mikhail Saulovich Arlazorov (1920-1980), a biographer of Konstantin Tsiolkovsky. In recounting some events from Tsiolkovsky's later years, Arlazorov mentioned that the late scientist had been invited to something called the All-Union Conference on the Use of Reactive Vehicles for the Study of the Upper Layers of the Atmosphere, held in Moscow in 1935. Tsiolkovsky had apparently declined the invitation due to health reasons; he actually died later that year. Arlazorov noted that "among [the list of those] who presented papers at the conference is the name of the chief designer of the Vostok spaceship". This was a key piece of the puzzle for the investigators at the Library of Congress, because the names of all the presenters of this conference had been published openly in the 1930s. So the researchers could start working through this list. But they needed more information to narrow down the search.

They found this in the essay when Arlazorov described a letter that Tsiolkovsky had received from the semi-governmental GIRD rocketry group. This organisation, the Group for the Study of Reactive Propulsion, was one of the first Soviet rocket research groups; by the early 1960s publications in the Soviet media had begun to discuss the work of GIRD. In his essay, Arlazorov described several artefacts from GIRD's work – letters, testimonials, memoirs, etc. One letter he quoted was rather interesting. He noted "[h]ere is a letter from the leaders of the Moscow GIRD", and then he provided a quote from the letter, presumably written by someone in GIRD: "Many qualified engineers are working with us, but the best of them is..." At this point Arlazorov censored the letter, saying that "here follows the name of the chief designer of the Vostok spaceship..." Obviously, Arlazorov was not allowed to print this name. He provided another clue:

> The future chief designer mailed a book to Kaluga [where Tsiolkovsky lived] but without his return address. "I do not know how to thank him for his kindness," wrote Tsiolkovskiy, "Thank him for me, if possible, or send me his address." [37]

Based on this information, the Library of Congress concluded that: (1) the Chief Designer had read his paper at the All-Union Conference in 1935; (2) he was the best engineer working at GIRD; and (3) he had sent his recent book to Tsiolkovsky in Kaluga. The last piece of information proved particularly useful. The Library of Congress found that only two major monographs were published in 1934-1935 by Soviet authors on the topic of rocket technology: M. K. Tikhonravov's *Raketnaya tekhnika* (*Rocket Technology*) and S. P. Korolev's *Raketnyy polet v stratosfere* (*Rocket Flight in the Stratosphere*). So one of them was quite likely the mysterious Chief Designer. But from Tsiolkovsky's letter it was clear that Tsiolkovsky did not know the author of the book personally, as implied by Tsiolkovsky's comment that he did not know the author's address. Yet, the Library of Congress also knew (from newspaper accounts from the 1930s) that Tsiolkovsky had actually *met* Tikhonravov and corresponded with him. Hence, by a process of elimination, they concluded that the Chief Designer *must* be this person "S. P. Korolev", who, it transpired, was also on the list of those presenting papers at the 1935 conference. They used several other

9.3: In Moscow recently with the Chief Designer's daughter, Natalya Koroleva.

published sources to corroborate this claim without inconsistencies. So in the end it was a Soviet book, passed by the censor, that gave the US its best clues to the actual identity of the mysterious Chief Designer – Sergei Korolev had been identified.

Published internally in May 1964, the Library of Congress study was available to the general public but was not widely distributed. It opened the door for much more widespread identification of Korolev. In November 1965, Theodore Shabad of the *New York Times* again openly identified Korolev and Glushko as the two leading chief designers [38]. It is not known if these major press reports were communicated back to Korolev himself. But just eight days before Korolev's death in January 1966, *Fortune* magazine, one of the most influential magazines in the United States, ran a story that identified him as the mysterious Soviet Chief Designer of Rocket-Space Systems [39]. Authored by journalist William Shelton, who would later write one of the first comprehensive books on the Soviet space programme, the article reinforced Korolev's role in the Soviet space programme [40].

So what do all these news items about Korolev cumulatively tell us? The most significant issue here is that *undoubtedly*, if he had not unexpectedly died during a medical procedure, his identity would have been widely known all over the world. There would have been more speculative articles and perhaps even questions posed directly to Soviet authorities about Korolev. And it is not unreasonable to argue that given all these sleuthing revelations, the Soviets themselves would have declassified his identity and he would have taken on an iconic public role similar to other giants of the Soviet weapons industry such as Andrey Tupolev and Igor Kurchatov. Alas, this was not to be. His death saved the authorities from having to confront this possibility. And a myth of sorts has been cultivated suggesting that the official Soviet announcement of Korolev's death in January 1966 was the defining moment in the West's recognition of his identity. But the evidence shows that Korolev was already a known name – not particularly famous – but known just the same. The veil had already been lifted by the time of his death.

In death and memory

When Sergei Pavlovich Korolev passed away on 14 January 1966, news of his death was reported widely in the Western press. Yet reports at the time still lacked certain essential pieces of information, and many Western newspapers did not immediately perceive the importance of his accomplishments. For example, the *New York Times* reported news of his passing on page 82 of the Sunday news edition, noting that he was a "leading Soviet space scientist" [41]. On being informed, James E. Webb, the NASA administrator, considered sending a message of condolence to Moscow but after discussions with several senior officials he decided not to do so [42].

But within days, the scope of Korolev's contributions became evident. In a major editorial in the *New York Times*, the editors noted that "death has finally declassified the role and identity of Academician Sergei P. Korolev, the man who provided the scientific and technical leadership of the Soviet rocket programme". They went on, "Korolev's rockets were powerful enough to send men into orbit and to put cameras into position to photograph the back side of the Moon. But they were too weak to break the chains of secrecy that denied him, while he lived, the world applause he deserved" [43].

In the first years after Korolev's death, Westerners discovered more about his life but almost all of it dated to *before* World War II and was based on the brief obituary published at the time of his death. The few pieces of information on his life after the war were connected with the dates of various awards and honours. To add insult to injury, even as late as 1968, some Western newspapers still said that Academician Leonid Sedov was the "father of Sputnik" [44].

Defector provides new details

A number of books on the history of the Soviet space programme were published in the United States and Britain in the late 1960s and early 1970s. In these books, the authors began to piece together a chronology of Korolev's life, although a lot of the information was still based on rumour and hearsay. Perhaps the most controversial aspect of Korolev's life was his incarceration as a prisoner of the Stalinist gulag system between 1938 and 1944. Details of his incarceration are now well-known but in the 1970s there was much that was uncertain. Analysts had only bits and pieces of information as evidence. For example, less than five months after Korolev's death, a Hungarian publication made the sensational claim that he had been in prison from

1940 to 1953, i.e., until Stalin's death. Days later this news made the pages of the *Washington Post* with the headline "Top Soviet Space Designer Worked in a Stalin Prison" [45].

Further details emerged in the late 1960s and early 1970s from a former Soviet journalist named "Leonid Vladimirov" who had defected to Great Britain in 1966. Like Korolev, Vladimirov, whose real name was Leonid Vladimirovich Finkelshtein (1924-), had spent time in the gulag; he was arrested while a student at the Moscow Aviation Institute in 1947 and spent six years in prison. Later, he became a staff writer for the popular science journal *Znaniye-sila* (*Knowledge is Power*) and met journalists who had access to the "inside" world of the Soviet space programme. After his defection "Vladimirov" wrote about Korolev's life (including his time in prison) in a number of publications. Finkelshtein's book *The Russian Space Bluff*, published in 1971, caused quite a sensation in the West [46]. There was much back-and-forth in the pages of the British magazine *Spaceflight* between those who found the book as credible and those who found it full of dubious claims. Hindsight and posterity have not been kind to *The Russian Space Bluff*. While it is true that it has some valuable insights (such as the now-accepted fact that the Voskhod was by-and-large a Vostok crammed with three cosmonauts), the book was also misleading in many ways and frequently full of inaccuracies; for example he called Soviet rocket designer Mikhail Yangel a German! Nevertheless, the book was very influential in the English-speaking world, and inspired others to undertake historical research on Korolev's years in the *sharaga* (often also called, *sharashka*) prison system.

Another book from the early 1970s that was smuggled out of the Soviet Union, claimed to be a memoir of "G. Ozerov" who had spent time in the *sharaga* prisons with Korolev. It was Ozerov's book that gave us one of the most famous alleged quotes from Korolev: "We will all vanish without a trace" [47]. Later, at the tail end of *glasnost*, it turned out that "Ozerov" was actually Leonid L'vovich Kerber (1903-1993), a deputy to famed Soviet aviation designer Andrey Tupolev who had indeed spent time in the camps with Korolev [48].

First Western 'biography'

With these and other works, the genie was out of the bottle. Western historians were able to combine snippets of information from many different sources – including the books of Mark Gallay, Roy Medvedev, and Aleksandr Solzhenitsyn, with rumours from other sources – to attempt to reconstruct Korolev's activities during the 1930s and 1940s [49].

The most important historical work in this regard was American journalist James Oberg's classic article 'Korolev and Khrushchev and Sputnik' in the British journal *Spaceflight* in 1978 [50]. Besides details on Korolev's incarceration, Oberg's article contained the first account of Korolev's activities in the postwar years – particularly his involvement in the development of the R-7. There were a few inconsistencies in the biography. For example, Oberg speculated that Korolev might have been arrested a second time in the late 1940s; he was not. In addition, although Oberg gave details of the infamous Nedelin disaster in 1960, he claimed that this took place during the launch of a Mars automatic interplanetary station as opposed to the attempted launch of an ICBM – later found to be Yangel's R-16 missile. Oberg used as a source for this the alleged memoirs of one Oleg Vladimirovich Penkovsky (1919-1963), a Soviet military intelligence officer who informed on the USSR to British and American authorities in the early 1960s and was later caught and executed by the Soviets [51]. Penkovsky provided tantalising details about the disaster, including the fact that the city of Dnepropetrovsk in Ukraine was in mourning afterwards; we now know, of course, that the R-16 ICBM was produced by the Yuzhnoye design bureau based in Dnepropetrovsk.

Despite some shortcomings, Oberg's article was the first substantive biography of Korolev published probably *anywhere* in the world; i.e., even including the Soviet Union. Oberg argued that Korolev was a pawn of politics, particularly of the whims of Nikita Khrushchev, and was forced to perform many space missions against his will. Oberg later expanded these observations into *Red Star in Orbit*, published in 1981. As I said above, the significance of this book cannot be overstated because it drew a large popular audience into the study of Soviet space history. There had, of course, been good books published in the West on the history of the Soviet space programme in the 1960s and 1970s. These included the works of Firmin J. Krieger (1909-), Albert Parry [née Paretzky] (1901-1992), Alfred J. Zaehringer (1925-2012), Martin Caidin (1927-1997), William R. Shelton, Michael Stoiko (1919-2010), Nicholas Daniloff (1934-), Piet Smolders (1940-), Peter N. James and Nicholas L. Johnson. But what distinguished Oberg's work was a certain flamboyance coupled with assiduous and exacting research. He was also not afraid to tell a good story – although not at the expense of the facts. One of Oberg's most striking conclusions was about Korolev's role in the Soviet space programme. He noted that "Korolev's premature death...may have been the most important contributing factor which prevented [a] cosmonaut lunar flight from occurring" [51]. In Oberg's imagination, the failure of the mythical lunar programme and the super booster programme was inextricably tied to Korolev's life and death.

REVEALING GLUSHKO

Whilst Korolev was clearly the central figure in the Western visualisation of the Soviet space programme, gradually, imperceptibly, through the 1970s and into the 1980s, it became apparent that there were others of equal importance whose names were unknown. Undoubtedly the most prominent among the others was Valentin Petrovich Glushko (1908-1989). Through the late 1960s, Glushko continued to publish under his assumed name of "Professor G. V. Petrovich" and he even edited the first major encyclopedia of spaceflight in 1968 [52]. For reasons that still remain unknown, in 1971, just prior to the first Salyut station missions, the Soviet censors decided to declassify his name in a dramatic manner. They not only identified him as the Chief Designer of Rocket Engines but also confirmed that "Professor G. V. Petrovich" had been a pseudonym that Glushko had used for many years. Soviet official sources naturally declined to explain why Glushko had needed a pseudonym for so long [53]. The first open interview with Glushko was published in the Moscow communist youth daily *Moskovskiy kosmomolets* in October 1972. In the interview he spoke at length about the future of chemical, nuclear, and electrical rocket propulsion [54].

Throughout the 1970s Glushko published many articles and was often interviewed by the Soviet media. In 1977 his early work was summarised in a 504-page volume *Put' v raketnoy tekhniki* (*The Path in Rocket Technology*), and this began the process of embellishing Glushko's contributions to the space programme at the expense of those of Korolev. As he took the helm of Korolev's old organisation (now known as NPO Energiya), Glushko began to position his contributions as equal to if not greater than those of Korolev. For example, as one of his first acts after taking over in 1974, he instructed the curators of Energiya's then highly restricted "display hall" to remove all traces of Korolev's handiwork (including the famous R-7 rocket that placed Sputnik into space) and to replace them with his own rocket engines. During the late 1970s and 1980s, Glushko sought to rewrite the official historical narrative in subtle ways that were not immediately noticed by Westerners; for example, book chapters on his own research preceded those on Korolev's research [55].

When did we first learn of the conflict between Korolev and Glushko? One would expect that this would have been revealed during the time of *glasnost* but, in fact, in the mid-1970s there were clues to the rift between the two giants of the Soviet space programme. In the smuggled memoirs of Nikita Khrushchev (published in English in 1974), the former Soviet leader noted cryptically that: "The principal designer of the [R-7] booster was Korolev's friend and collaborator, whose name I forget. The best booster in the world won't make a broomstick fly. So while Korolev designed the rocket, his colleague [designed] the engine. They made an excellent team. Unfortunately, they split up later. I was very upset and did everything to patch up their friendship, but all my efforts were in vain [56]."

When the unedited portion of this passage was finally published in 1990, we found some added details: "...differences of opinion started to pull [Korolev and Glushko] apart and the two of them couldn't stand to work together. I even invited them to my dacha with their wives. I wanted them to make peace with each other, so that they could devote more of their knowledge to the good of the country, rather than dissipate their energy on fights over details. It seemed to me that they were both talented, each in his own field. But nothing came of our meeting. Later Korolev broke all ties with Glushko [57]."

Officially revealing Glushko's identity in 1971, while he was still alive and very much active, was unprecedented. It was a striking example of the enormous power which the rocket engine designer wielded, a level of influence matched by few of his contemporaries. The identities of only a very small group of designers in the Soviet defence industry were revealed during their lifetimes. The usual custom was for death to "reveal" a designer's identity and work [58]. This is how we *officially* learned the names of Mikhail Yangel and Vladimir Chelomey.

Korolev's successor?

The way the names of these two men came to light was hostage to a fundamental misunderstanding among Western sleuths, with analysts assuming that there was a single and massive research and development organisation that had been headed by Korolev. A natural assumption was that after his death another Chief Designer took over. The most likely contender for a "successor" was Mikhail Kuzmich Yangel (1911-1971), the Soviet rocket designer who proved to have been responsible for several generations of strategic ICBMs, space launch vehicles, and automated military and scientific satellites. In June 1966, just five months after Korolev's death, the *New York Times* ran a short piece on Yangel by Theodore Shabad – the same journalist who had correctly identified Korolev a few years earlier. He wrote, "A 54-year-old Ukrainian engineer, who has recently been advanced to high position in the Kremlin hierarchy, has been tentatively identified as the new scientific head of the Soviet Union's secrecy-shrouded space programme. The promotion of the Siberian-born scientist, Mikhail K. Yangel, to a public position is believed to reflect a high-level political decision to give a few leading space technicians, usually cloaked in anonymity until their death, general recognition during their lifetime while still avoiding open identification of their work [59]."

Shabad's work was based on his investigation of the published lists of names of people who had been "promoted" to Candidate Membership of the ruling Central Committee of the Communist Party. Such an honour was typically reserved for the most influential citizens of Soviet civil society. While it was a rank that was largely honorific ("full" members of the Central Committee were more likely to have true power in the Party hierarchy), by scrutinising such lists – in addition to the order of signatures on obituaries of famous Soviet individuals, published lists of members of the rubber stamp Supreme Soviet, the faces of people who showed up at parades in Red Square, the signatures on articles in *Pravda*, etc. – Western observers were able to determine the intricacies of whose fortunes rose or fell in the halls of power. This was actually a fairly well-established field in the West known as "Kremlinology".

In his article, Shabad also identified Valentin Glushko, Nikolay Pilyugin, and Grigoriy Kisun'ko as potentially important missile designers [60]. There was little further information on Yangel until his death in 1971 when his identity was officially revealed, albeit in rather vague terms. His obituary simply acknowledged that he was "an outstanding scholar and designer in the field of rocket and space technology". Western media outlets continued to tout him as Korolev's *successor* rather than the head of an entirely different organisation [61].

A leading missile-man

The name of Vladimir Nikolayevich Chelomey (1914-1984) had been mentioned by the spy Oleg Penkovsky in *The Penkovsky Papers* (1965) and also by the defector Finkelshtein in his 1971 book *The Russian Space Bluff* (although again with many inaccuracies), but a more substantive identification that Chelomey was a major chief designer in the Soviet space programme came from Nicholas Daniloff in his book

The Kremlin and the Cosmos in 1972 [62]. As with Yangel, Chelomey's promotion within the Communist Party hierarchy in 1974 prompted Theodore Shabad to claim that Chelomey was "the new head of Russia's secrecy-shrouded space programme... [a] job that was previously held by Mikhail K. Yangel [63]."

Bits and pieces of information about Chelomey continued to trickle out through the 1970s and early 1980s, but it was not until his death in December 1984 that we got the first concrete details about his life; his obituary said he was an "outstanding designer of Soviet rocket technology and flying vehicles" [64]. Soon after, in 1985, articles began to appear linking Chelomey with the Proton launch vehicle. Over the next few years a Soviet journalist named Valeriy Yevgen'yevich Rodikov published several articles which revealed details of Chelomey's colourful and rich career as a designer of cruise missiles, ICBMs, space launch vehicles, manned spacecraft, and military satellites [65]. Most of this information quickly trickled out to the West and appeared in *Spaceflight* magazine or various French publications, but the true extent of Chelomey's massive contributions to the Soviet space programme was not known in the West until the 1990s helped partly by the writings of Sergey Khrushchev, the son of the Soviet leader, who emigrated to the United States.

The younger Khrushchev had worked for Chelomey between 1958 and 1968 as a guidance systems engineer. In the 1990s Khrushchev became a widely sought-after interview subject for many Western analysts – I myself interviewed him in 1996 for

9.4: Interviewing Vladimir Polyachenko in 2007. In the background is Timofey Prygichev, a space sleuth based in St. Petersburg.

my first book. He later invited me to fact-check his English-language memoirs (and was kind enough to thank me in the acknowledgments) which were finally published in 2000 [66]. I also developed a friendship with Vladimir Abramovich Polyachenko (1929-), who Chelomey had appointed "lead designer" for several key programmes, including the IS anti-satellite project and the Almaz space station. I helped Vladimir to publish a portion of his recollections about the competition between Korolev and Chelomey in *Spaceflight* magazine in 2011 [67]. My hope is that working together with Vladimir I will eventually be able to publish a more comprehensive history of the Almaz space station programme [68].

Unlike journalists or analysts who worked only with public information, Western intelligence agencies appear to have had a very good sense of the entire design bureau system by the late 1970s. A lengthy National Intelligence Estimate (NIE) issued by the CIA in 1980 on 'Soviet Military Capabilities and Intentions in Space' correctly showed a chart of Soviet design bureaus involved in the space programme; including those of Valentin Glushko, Mikhail Reshetnev, Vladimir Utkin, Sergey Kryukov, and Vladimir Chelomey [69]. These were some of the principal chief designers at the time, and it is a testament to the CIA's access to (probably) human intelligence information that this information – all correct – was known to them.

Back from obscurity

Perhaps the most enigmatic trajectory of a Soviet chief designer was that of Vasily Pavlovich Mishin (1917-2001) who succeeded Korolev at the OKB-1 design bureau. Mishin was hardly known in the West until his revelations in 1989 about the manned lunar landing programme. Yet even Mishin's name was linked to the Soviet space programme in the early 1970s while he was still a Chief Designer.

Like several other designers, Mishin wrote or edited many arcane mathematical textbooks under his own name in the 1960s, 1970s and 1980s. These were available at such places as the Library of Congress in Washington, D.C., but few suspected the author of having anything to do with the Soviet space programme [70]. On the other hand, when writing on space topics he used the pseudonym "M. Vasil'yev". During his tenure as Chief Designer, he also edited at least two important books on the Soviet space programme under this pseudonym – *Steps to the Stars* (1972) and *Salyut In Orbit* (1973) – both of which were later translated by NASA into English in the run up to the Apollo-Soyuz Test Project [71]. He also wrote major articles (also under his pseudonym) for newspapers such as *Pravda, Izvestia,* and *Krasnaya zvezda* [72].

French sleuths played an important role, and in 1972 journalist Pierre Dumas was able to link Mishin's name with the Soviet space programme in connection with an article on Soviet plans for future manned Mars expeditions [73]. Based on this article and some other unconnected clues, several months later a Ukrainian émigré for the first time argued that Mishin was the mysterious Chief Designer of the Soviet space programme. But as he published this in an obscure émigré journal based in the United States, few paid any attention [74]. In 1974 French analyst Claude Wachtel listed an "M. P. Vassiliev" in Volume 11 of the French *Cosmos Encyclopedia*. Then in 1977 Christian Lardier explicitly identified "V. P. Michine" in a French popular science book [75]. A Soviet defector who worked at the old Korolev design bureau added some tantalising details in a quasi-memoir that he published in the United States in 1982. The author, a "Victor Yevsikov", correctly claimed that Vasily Mishin had succeeded Korolev as Chief Designer, and described the loss of prestige that came with Korolev's death [76]. Despite these prominent claims, most Western analysts remained unaware of Mishin until well into the 1980s.

French analyst Claude Wachtel was the first in the West to emphatically claim that Vasily Mishin was the successor to Korolev. In a landmark article published in 1985 based on a paper he presented at the annual Soviet Forum of the British Interplanetary Society in London in 1983, Wachtel actually published a photograph of Mishin [77]. Remarkably, by the time that Wachtel's article was published, Mishin's position and importance in the Soviet space programme had still not been acknowledged *within* the Soviet Union. In fact, when Glushko took over from Mishin at NPO Energiya in 1974, he had made sure that Mishin's name was white-washed out of history. During his years of "banishment" as a professor at the Moscow Aviation Institute, Mishin quietly worked on a number of important historical projects, including collecting the works of rocketry pioneer Fridrikh Tsander and editing a classic volume of Korolev's works that was published in 1980 and revealed an astounding number of secrets about the Soviet space programme. Discerning owners of *The Creative Legacy of Sergei Pavlovich Korolev* will see that Mishin's name was one of the many listed as part of the editorial council for the book [78]. He also co-wrote the main introduction to the book, although he was only identified as an "Academician". The Soviet media only began to fully acknowledge Mishin's role in the Soviet space programme in 1987, in connection with the twentieth anniversary of Sputnik [79]. The door was pushed wide open with the famous interview Mishin gave to the newspaper *Pravda* in 1989 on the failed Soviet manned lunar programme [80]. Within a few months his name was on the lips of all space sleuths. And then in late 1990 he published a tour de force history of the Soviet manned lunar project in which he provided a slew of previously unknown details on the project [81].

The Mishin diaries

During his time as Korolev's deputy (and later as his successor) Mishin kept daily office notes about his activities. In the chaos after the collapse of the Soviet Union, Mishin offered these 32 notebooks to be sold at a Sotheby's auction in late 1993. The Perot Foundation (named after the American businessman and entrepreneur H. Ross Perot) bought them, allegedly for $190,000 [82]. Copies of the diaries, albeit of poor quality, have found their way to at least a couple of different archives in the United States and I was able to make use of some of them for my book *Challenge to Apollo*. Along with fellow Soviet space sleuth Peter Gorin, I was invited to work on a portion of these copies to explore the option of releasing them to the public.

I had made contact with Peter in 1993 or so, as a result of our common interest in uncovering the secrets of the Soviet space programme. Peter, a Russian citizen, had

9.5: Posing with a once-secret Soviet lunar lander at Mishin's Moscow Aviation Institute.

9.6: A copy of *Challenge to Apollo* was presented to Mishin shortly before his death by his former MAI student Dmitry Payson.

emigrated to the United States with his family in 1990. He was a political scientist by training with the equivalent of a doctorate from a prominent Moscow university. His knowledge of the inner details of the Soviet space programme was encyclopedic. In the 1990s, he published some ground-breaking articles on the history of the Soviet photo-reconnaissance programme, in particular about the Zenit and Yantar series of satellites, and he worked for a long time on the history of the Soviet manned lunar programme [83]. In the early 2000s, Peter and I began to work on Mishin's diaries, but he found it difficult to devote his full attention due to health-related problems and struggles involving his naturalisation process in the United States. Sadly he suddenly passed away on 16 January 2009 in Norfolk, Virginia, 56 years of age [84].

Peter's untimely death, together with various logistical challenges, have impeded work on the Mishin diaries project. One hopes that one day these priceless notebooks will see the light. Although I was never fortunate enough to meet Mishin, through an intermediary I was able to send a copy of my *Challenge to Apollo* to him. One of my most prized possessions is a picture of Mishin holding a copy of the book, taken only a few months before his death on 10 October 2001. In January 2012, with NASA's Chief Historian Bill Barry, I was given a private tour of Mishin's former MAI office and it was humbling to be in the same room where the leading architect of the Soviet manned lunar programme had worked [85].

INTO THE ARCHIVES

When revelations about each of the prominent Soviet designers appeared in the late 1980s and early 1990s, Western sleuths were employing a combination of sources to reconstruct the hidden stories of the Soviet space programme: articles in the Russian media; official Soviet-Russian books and press releases (often containing revealing photographs); analyses of orbital behaviour; rumour and speculation; personal and uncensored interviews with designers and cosmonauts; and declassified intelligence documents from the CIA. But by about the year 2000, one further source of analysis became available to Westerners: the actual archival documents. I was fortunate to be amongst the first Westerners to work with space-related archival material held in the Russian archives. This has played a crucial role in my more recent work, including highlighting further and deeper secrets about the men who were the architects of the Soviet space programme.

The very first missile/space publications based on archival documents were works on the German contribution to the Soviet missile industry in the late 1940s. German historians Matthias Uhl and Christoph Mick published stellar works based on a deep mining of such sources, most of them at the Russian State Archive of the Economy (RGAE) in Moscow [86]. Starting in 2002 I worked for many months at RGAE and other archives, including the State Archive of the Russian Federation (GARF), the Russian State Military Archive (RGVA), and the Archive of the Russian Academy of Sciences (ARAN). This work served as the foundation for my recent book, *The Red Rockets' Glare: Spaceflight and the Soviet Imagination, 1857-1957*. Published by Cambridge University Press in 2010, the book is the very first analysis of the early history of the Soviet missile and space programme based almost entirely on archival sources.

I am often asked what it was like to work in these archives, and about the kind of documents that are available. The actual experience of archival research in Moscow bears only a passing resemblance to similar work in the Western world. Yes, there are rudimentary finding aids. Yes, there are reading rooms. But there are also substantial differences. For a start, there are many more security restrictions. One also has to deal with a Byzantine bureaucracy. Finally, the degree of access one gets to certain files is often a function of personal relationships or the whims of archivists. Despite these idiosyncrasies, the prize at the end of the process can be incredibly fulfilling. Below, I provide some brief examples of the kind of documents I collected on the work of Korolev, Glushko, Yangel, Chelomey and Mishin.

One of the most prized finds in my archival work was by accident. For some time, I had been obsessed with the Soviet manned lunar programme. One key aspect that remained clouded in my mind was the Soviet decision to go to the Moon: how, why,

9.7: At the entrance to the Russian State Archive of the Economy in 2006.

and when did the Soviets decide to compete with Apollo? What precisely prompted them to mount a challenge to Apollo? And what role did the five men listed above play in this decision? Inspired by a classic work of American space history, John M. Logsdon's magisterial *The Decision to Go to the Moon* analysing John F. Kennedy's famous commitment made in May 1961 to land a man on the Moon before the decade was out, I wrote about the Soviet side in a two-part article in *Spaceflight* in 1998 [87]. But without actual documents in my hand, I still felt uncertain about my conclusions.

In 2002 I stumbled upon what I thought of as the 'holy grail'. While working at RGAE, I had continually butted heads with archivists about the absence of certain papers concerning the Soviet defence industry (which oversaw the space programme). It had taken me a while to figure out that such papers were actually not stored in the main RGAE building on *ulitsa Bol'shaya Pirogovskaya* near the Frunzenskaya metro station in Moscow. I slowly realised that there was a *odtel spetsfondov* ("Department of Special Funds") at an entirely different location which was unlisted in any archive finding aid. I would need special permission to go there, so I spoke personally with the deputy director of the archive, who was sympathetic but suspicious. Finding my way to the rather unremarkable building near the Kaluzhskaya metro station that held this "Special Fund", I found a treasure trove of

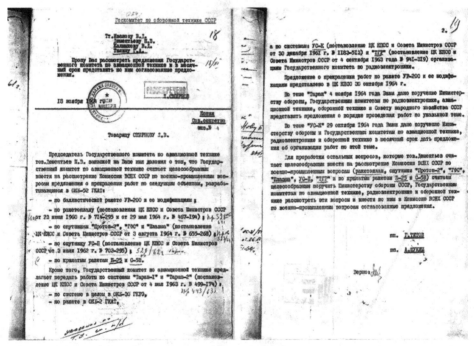

9.8: This two-page document was prepared after the 'fall' of Nikita Khrushchev and proposes the cancellation of several Chelomey projects.

documents. Here, in 2003, I stumbled, quite accidentally, upon what was clearly a draft version of the *actual* decree from 1964 that committed the Soviet Union to a manned landing on the Moon. I published my analysis of this document in a lengthy article in *Spaceflight* in 2004 [88].

Among the many revelations from this decree were details of programmes that were part of the effort that fell by the wayside later (such as Chelomey's Oriented Lunar Satellite), several abandoned scientific satellite (Protsion, Plazma, GFS, etc.), and the projected cost of a manned lunar landing (2.1 billion rubles). As part of this research I also identified the precise date on which Korolev met with Khrushchev to discuss the terms of the decision to go to the Moon (17 July 1964). Finally, the fact that both Korolev and Chelomey signed the document suggests that there was a level of détente between the two competitors. I am proud to say that I pre-empted Russian scholars by a few years – the same decree finally appeared in print in a Russian book in 2008 [89].

My sleuthing comes full-circle

Around the time that I published this decree, I had become well-acquainted with two persons who had been close to Korolev: his daughter Natalya Sergeyevna Koroleva (1935-) and one of his leading deputies, Boris Yevseyevich Chertok (1912-2011). I had gotten to know the latter while helping him to translate and edit his memoirs for an English-speaking audience [90]. Chertok in turn introduced me to Koroleva, who has been a very generous and kind host to me on each of my trips to Moscow. As I got to know them better, I had to 'turn off' my sleuthing mode and resist the urge to ask them questions at every opportunity about some arcane topic or other. But it was heartening and fulfilling to see their generosity and openness to Western researchers. In Koroleva's home there is a small "museum" dedicated to her father. The artefacts include Korolev's various awards, letters, models, and possessions. On a bookshelf, among various prized Russian books that Korolev himself owned, she has a copy of the late Rex Hall and David Shayler's *The Rocket Men* and James Harford's *Korolev* [91]. She considers them important contributions to under-standing her father's life. I was also grateful to see that in the introduction that Chertok wrote to Koroleva's 3-volume work *Otets* (*Father*), he mentioned my book *Challenge to Apollo* as one of the most important Western works on Korolev and the Soviet space programme [92]. Since this was before I actually came to know Chertok personally, I felt particularly honoured. In an odd way, Western sleuthing had come full-circle, with the Russians themselves now mentioning our work as standards. This is enormously gratifying.

In the Moscow archives I was also able to find many other key documents about the work of the leading chief designers, examples of which include: the interrogation records of Glushko in early 1938 before his arrest and incarceration; documents from 1945 proposing that Glushko and *not* Korolev should be assigned "Chief Designer" of the Soviet ballistic missile programme; letters from Korolev to his (second) wife describing Glushko's ill-behaviour towards him; the annual reports of the Chelomey design bureau from the early 1960s which provide many unknown

9.9: Sitting with Boris Chertok and Vladimir Syromyatnikov, the designer of the Soviet docking system, on "Korolev's bench" in May 2006.

details about the two Polet satellites; letters detailing the "collapse" of the Chelomey empire after the fall of Nikita Khrushchev in 1964 when several of his projects, including the UR-200, GFS, and Plazma were cancelled; and documents showing how Chelomey justified the UR-700 project as a competitor to Mishin's N-1.

The archives contain many documents on the strategies the Soviets employed to maintain secrecy, which I used as the basis for a recently published essay 'Cosmic Contradictions: Popular Enthusiasm and Secrecy in the Soviet Space Programme' [93]. Some of these documents point to the use of bland designations for spacecraft.

Soviet attempt to confuse?

One of the most tantalising finds was a set of documents detailing how the Soviets sought to *confuse* Western space sleuths. By the mid-1960s, Western intelligence analysts (as well as others such as Geoff Perry and Charles Sheldon) had begun to realise that the 'Kosmos' series of satellites were largely military in nature. There were also numerous failures hidden under that catchall cover name. In an effort to further confuse Westerners, in 1965 Soviet space policy-makers proposed adding a new class 'Zarya' (Dawn) satellites which would, like Kosmos, include military, non-military, precursor test flights, and failed missions [94]. Satellites were to be split

9.10: Pictured at the Moscow office of *Novosti kosmonavtiki* in January 2012 are: (L-R) Igor Afanasyev, Igor Lissov, Igor Marinin, Asif Siddiqi and NASA's Chief Historian Bill Barry.

randomly between the Kosmos and Zarya designations. The plan was shelved but it is not clear why; perhaps a hapless bureaucrat pointed out that having the Kosmos and Zarya designations would not only confuse foreigners, it would probably also confuse the Soviets, so perhaps it would be wise to stick with the single Kosmos designation!

To end the section on archival research on the Soviet space programme, it is only appropriate to also acknowledge the work of native Russians in ferreting out original sources. The monthly journal *Novosti kosmonavtiki* (*News of Cosmonautics*) has been at the forefront in uncovering formerly secret information for more than two decades. Its noteworthy sleuths have included Igor Afanasyev, Vladimir Agapov, Konstantin Lantratov, Igor Lissov, Igor Marinin, Sergey Shamsutdinov, and the late Maxim Tarasenko (1962-1999). I would also include two professional historians. The first was Georgiy Stepanovich Vetrov (1918-1997) who was responsible for decades of historical research in the archives of NPO (later RKK) Energiya; it was because of his diligent work that we are fortunate to have available the 716-page volume of declassified documents about Korolev's work published in 1998 as *S. P. Korolev i ego delo* (*S. P. Korolev and His Affairs*) [95]. Vetrov, a former engineer under Korolev, worked on a number of other similar volumes, most of which unfortunately remain unpublished; his untimely death having stalled this valuable historical work. In the mid-1990s, I was fortunate to have corresponded with Vetrov via the late Maxim Tarasenko, the young author of the first Russian-language book on the history of Soviet military programmes [96]. On the military side, another leading figure has been military historian Vladimir Ivanovich Ivkin (1958-) who has helped to declassify hundreds of documents on the origins of the Soviet ICBM programmes. Since 1993 he has served as a historian at the Russian Strategic Rocket Forces. Perhaps the most important of Ivkin's works was the nearly 1,200-page-long volume of declassified documents on the Soviet missile programme published in 2010 under the title *Zadacha osoboy gosudarstvennoy vazhnosti* (*A Goal of Special State Importance*) [97].

CONCLUSIONS

In this chapter I have provided an account of how prominent Soviet space designers' names became public knowledge in both the Soviet Union and the West. In all four cases – Sergei Korolev, Valentin Glushko, Mikhail Yangel and Vasily Mishin – they were identified in some fashion by Western analysts prior to their deaths. I think the most striking revelation here is about Korolev: a number of Western journalists and analysts had conclusively identified him during his lifetime as the enigmatic "Chief Designer" of the Soviet space programme. But it is still common for journalists and historians to write that Korolev's identity was unknown to the general public during his lifetime. True, his name was a state secret and few in the Soviet Union had heard of him. But the evidence used here shows that in the West, diligent researchers had already ferreted out the truth.

Historians need to dispense with the myth that Korolev was a complete unknown during his life. Undoubtedly, had Korolev lived past January 1966 his real job would have become common knowledge, and space books written in the late 1960s might have started with a biography of him. One wonders what Korolev would have thought of that. On the other hand, it is also clear that the Soviet government took great pains to hide the names of their leading space scientists and engineers. Although Korolev's name might have been known in the West, Westerners knew few details of his life or his actual accomplishments. As with the lives of other important designers – such as Glushko, Chelomey or Mishin – it took the intrepid work of Western sleuths (such as Dennis Newkirk and Peter Gorin) to uncover the true details of their amazing lives. The next step, I believe, will be for sleuths to get direct access to archival documents and peel off yet another layer of secrets from the amazing history of the Soviet space programme.

References

1. *Peresadka na orbite* / *Transfer in Orbit* (Moscow: Novosti, 1969). The text was both in English and Russian.
2. G. V. Petrovich, ed., *The Soviet Encyclopedia of Space Flight* (Moscow: Mir Publishers, 1969).
3. Reginald Turnill, *The Observer's Book of Manned Spaceflight* (London: F. Warne, 1975). This was a revised edition of a book that originally came out in 1972. A third edition came out in 1978.
4. James E. Oberg, *Red Star in Orbit* (New York: Random House, 1981).
5. By 1985, the following editions had been published: *Soviet Space Programs* (published in 1962), *Soviet Space Programs, 1962-65* (1966), *Soviet Space Programs, 1966-70* (1971), *Soviet Space Programs, 1971-75*, 2 vols. (1976), and *Soviet Space Programs, 1976-80*, 3 vols. (1982, 1984, and 1985).
6. Nicholas L. Johnson, *The Soviet Reach for the Moon* (Washington, DC: Cosmos Books, 1994). A second edition was published in 1995.
7. Asif A. Siddiqi, "Soviet Space Programmeme – Organisational Structure in the 1960s." *Spaceflight* 36 (August 1994): 283-86 and 36 (September 1994): 317-20; Asif A. Siddiqi, "Mourning Star: The Nedelin Disaster." *Quest* 3, no. 4 (1994): 38-47.
8. Dennis Newkirk, *Almanac of Soviet Manned Space Flight* (Houston: Gulf Publishing Company, 1990). See also his "Soviet Space Planes," *Spaceflight* 32 (October 1990): 350-355.
9. Asif Siddiqi and Dennisk Newkirk, "The FGB Module of the International Space Station Alpha: A Historical Overview of its Lineage and Organizational Origins," Paper presented at the annual meeting of the Society for the History of Technology (SHOT), Charlottesville, VA, October 21, 1995.
10. Asif A. Siddiqi, *Challenge to Apollo: The Soviet Union and the Space Race, 1945-1974* (Washington, DC: NASA, 2000). This book was later republished in paperback in two separate volumes: *Sputnik and the Soviet Space Challenge* (Gainesville, Fla.: University Press of Florida, 2003) and *The Soviet Space Race with Apollo* (Gainesville, Fla.: University Press of Florida, 2003).
11. "Dennis Ray Newkirk," http://www.davenportfamily.com/UserFiles/Tools/Obituaries/showObit.cfm?obitID = 1061&keepThis = true. (accessed May 31, 2012).
12. Harry Schwartz, "Soviet Reticent on Space Chiefs," *New York Times*, October 5, 1959, p. 16. Khrushchev made the speech on July 9, 1958 in Bitterfield, East Germany.
13. There is, in fact, a whole book in the Russian language devoted to his contributions to aeronautics. See G. S. Vetrov, *S. P. Korolev i aviatsii. Idei. Proyekty. Konstruktsii* [*S. P. Korolev and Aviation. Ideas. Projects. Designs*] (Moscow: Nauka, 1988).
14. See *Krasnaya zvezda*, October 24, 1930 and November 10, 1930.
15. *Vestnik vozdushnogo flota* no. 2 (1931). Korolev was also mentioned in journals such as *Samolet* (*Airplane*) and *Nauka i tekhniki* (*Science and Technology*) and

other newspapers such as *Izvestiya* (*News*) and *Vechernyaya moskva* (*Evening Moscow*).

16. S. P. Korolev, *Raketnyy polet v stratosfere* [*Rocket Flight into the Stratosphere*] (Moscow: Gosudarstvennoye voyennoye izdatel'stvo, 1934).

17. Reviews were published in the newspaper *Za rulem* (March 21, 1935) and the journal *Kniga i proletarskaya revolyutsiya* (nos. 11-12, 1935, pp. 145-150).

18. S. P. Korolev, "K zavoyevaniyu stratosferu" ["To Master the Stratosphere"], *Tekhnicheskaya kniga* no 6 (1937): 98-99.

19. For details about his arrest and incarceration, see Asif A. Siddiqi, *The Red Rockets' Glare: Spaceflight and the Soviet Imagination, 1857-1957* (New York: Cambridge University Press, 2010), Chapter 5.

20. S. Korolev, "Osnvopolozhnik raketnoy tekhniki" ["Founder of Rocket Technology"], *Pravda*, September 17, 1957.

21. Asif A. Siddiqi, "Germans in Russia: Cold War, Technology Transfer, and National Identity" in *Osiris, 2nd Series, Vol. 24 (Science and National Identity)*, eds. Carol E. Harrison and Ann Johnson (Chicago: University of Chicago Press, 2009), pp. 120-143.

22. Irmgard Gröttrup, *Rocket Wife* (London: Andre Deutch, 1969). See also Gerald Schroder, "How Russian Engineering Looked To a Captured German Scientist," *Aviation Week*, May 9, 1955, pp. 27-34.

23. CIA, *A Summary of Soviet Guided Missile Intelligence*, US/UK GM M-52, July 20, 1953, p. L45.

24. CIA, Office of Scientific Intelligence, *Scientific Research Institute and Experimental Factory 88 for Guided Missile Development, Moskva/Kaliningrad*, OSI-C-RA/60-2, Scientific Intelligence Research Aid, March 4, 1960, pp. 11, 29.

25. CIA, *National Intelligence Estimate 11-5-61: Soviet Technical Capabilities in Guided Missiles and Space Vehicles*, Washington, D.C., April 25, 1961, p. 50.

26. V. A. Yegorov, "Iz vospominaniy o M. I. Lidovye" ["From Recollections about M. I. Lidov"], *Kosmicheskiye issledovaniya* [*Space Research*] no. 5 (2001): 451-453.

27. Yaroslav Golovanov, *Korolev: fakty i mify* [*Korolev: Facts and Myths*] (Moscow: Nauka, 1994), p. 553.

28. For the final published version, see Christian Lardier, "Soviet Space Designers When They Were Secret" in *History of Rocketry and Astronautics, AAS History Series, Vol. 25*, eds. Herve Moulin and Donald C. Elder (Novato, CA: Univelt, 2003), pp. 319-334.

29. Korolev's pieces (under the pseudonym "K. Sergeyev") were published in *Pravda* on December 10, 1957, November 10, 1960, October 14, 1961, December 31, 1961, January 1, 1964, January 1, 1965, and January 1, 1966.

30. G. V. Petrovich, "Vlasteliny ognennogo vodopada" ["Masters of Fire Falls"], *Komsomol'skaya pravda*, August 14 and 15, 1962, p. 2 (both issues).

31. Documents from Russian archives now confirm this aspect of Tokaev's mission. See the several dozen pages of documents stored in the Russian State Archive of the Economy (RGAE), *fond* [fund] 8044, *opis'* [inventory] 1, *delo* [file] 1647, *listov* [pages] 2-169.

32. " 'Agent' for Stalin Flees Soviet Zone," *New York Times*, September 1, 1948, p. 4; "TASS Writer Calls Tokayev 'Traitor'," *New York Times*, September 7, 1948, p. 8. Tokaty died in Cheam, Surrey on November 23, 2003.

33. For example, he claimed that "Professor G. Petrovich" was a real person and not a pseudonym for Valentin Glushko. Tokaty published two books on his experiences as a Soviet scientist, each with slightly different stories about the Sänger episode. See G. A. Tokaev, *Stalin Means War* (London: George Weidenfeld & Nicolson Ltd., 1951); G. A. Tokaev, *Comrade X* (London: The Harvill Press, 1956).

34. The text of his lecture was published in two places: G. A. Tokaty, "Soviet Space Technology," *Spaceflight* 63 (1963): 58-64; G. A. Tokaty, "Soviet Rocket Technology," *Technology and Culture* 4 no. 4 (1963): 515-528.

35. Theodore Shabad, "Soviet Space Planners' Identity Believed Known," *New York Times*, November 12, 1963, p. 2. For background to this story, see also Nicholas Daniloff, *The Kremlin and the Cosmos* (New York: Alfred A. Knopf, 1972), pp. 67-70.

36. Aerospace Information Division, Library of Congress, *Top Personalities in the Soviet Space Programme* (Washington, DC: Library of Congress, May 26, 1964).

37. M. Arlazorov, "Vse li my znayem o tsiolkovsom?" in *Nashi kosmicheskaya puti*, eds., S. V. Kurlyandskiya and N. Ts. Stepanyan (Moscow: Sovetskya rossiya, 1962).

38. Theodore Shabad, "Soviet Lifts Edge of Rocket Shroud," *New York Times*, November 7, 1965, p. 16.

39. William Shelton, "The Russians Mean to Win the Space Race," *Fortune* Vol. 73 no. 2 (February 1966): 140-143, 174-185.

40. Shelton's book was published in 1968: see *Soviet Space Exploration: The First Decade* (New York: Washington Square Press, 1968).

41. "Sergei P. Korolev Is Dead at 59; Leading Soviet Space Scientist," *New York Times*, January 16, 1966, p. 82. For other obituaries, see "Russians' Space Ship Designer Korolev Dies," *Washington Post*, January 16, 1966, p. 4; "Sergei Pavlovich Korolev: He Put Red Flag on the Moon," *New York Herald Tribune*, January 16, 1966, p. 7.

42. Webb to Frutkin, January 15, 1966, Biographical Files for Sergei Korolev, NASA History Division, Washington, DC.

43. "The Secret Scientist," *New York Times*, January 20, 1966, p. 32.

44. "Soviet Scientists Hail Apollo Flight," *New York Times*, December 27, 1968, p. 21.

45. Stephen S. Rosenfeld, "Top Soviet Space Designer Worked in a Stalin Prison," *Washington Post*, June 16, 1966, p. A27.

46. Leonid Vladimirov, *The Russian Space Bluff: The Inside Story of the Soviet Drive to the Moon* (New York: The Dial Press, 1971). See also L. Vladimirov, "From Sputnik to Apollo" (in Russian), *Posev* (September 1969): 47-51.

47. G. Ozerov, *Tupolevskaya sharaga* (Beograd: M. Cudina & S. Masic, 1971).

48. L. L. Kerber, *Stalin's Aviation Gulag: A Memoir of Andrei Tupolev and the Purge Era* (Washington, DC: Smithsonian Institution Press, 1996).

49. These authors produced important works that shed light on Korolev's time in prison. Mark Gallay (1914-1998), the famous Soviet test pilot who Korolev invited to witness Yuriy Gagarin's launch in 1961, described Korolev's time in prison in cryptic terms in his memoirs. See M. K. Gallay, *Ispytano v nebe* [*Testing in the Heavens*] (Moscow: Molodaya gvardiya, 1963). In his magisterial work on the history of Stalinism, Roy Medvedev (1925-) briefly mentioned Korolev's incarceration. See his *Let History Judge: The Origin and Consequences of Stalinism* (New York: Alfred A. Knopf, 1972). Famous Soviet dissident Aleksandr Solzhenitsyn (1918-2008) wrote about the *sharaga* camps, albeit in semi-fictional setting, in his famous novel *The First Circle* (New York: Harper & Row, 1968).

50. James E. Oberg, "Korolev and Khrushchev and Sputnik," *Spaceflight* (1978): 144-50. A prior version appeared as "Korolev," *Space World* (May 1974): 17-24.

50. See Oleg Penkovsky, *The Penkovsky Papers: The Russian Who Spied for the West* (New York: Doubleday, 1966).

51. Oberg, *Red Star in Orbit*, p. 118.

52. G. Petrovich, ed., *Malen'kaya entsiklopediya: kosmonavtika* [*Little Encyclopedia: Cosmonautics*] (Moscow: Sovetskaya entsiklopediya, 1968).

53. "Soviet Space Chief Identified as Editor of an Encyclopedia," *New York Times*, March 19, 1971, p. 24.

54. Theodore Shabad, "Russian Predicts 3-Engine Rockets," *New York Times*, October 16, 1972, p. 11.

55. For details, see Asif A. Siddiqi, "Privatising memory: the Soviet space programme through museums and memoirs" in *Showcasing Space*, eds. Michael Collins and Douglas Millard (London: The Science Museum, 2004), pp. 98-115.

56. Nikita Khrushchev, *Khrushchev Remembers: The Last Testament* (Boston: Little, Brown & Co., 1974), pp. 46-47.

57. Nikita Khrushchev, *Khrushchev Remembers: The Glasnost Tapes* (Boston: Little, Brown & Co., 1990), p. 186.

58. Designers whose names were revealed during their lifetimes included K. D. Bushuyev, K. P. Feoktistov, N. A. Pilyugin, and B. V. Raushenbakh,

59. Theodore Shabad, "Soviet Shift Hints New Space Chief," *New York Times*, June 26, 1966, p. 3.

60. Kisun'ko was the chief designer of the early Soviet anti-ballistic missile systems (System A and A-35M) at the giant KB-1 organization based in Moscow.

61. "Mikhail Yangel, Soviet Space Aide," *New York Times*, October 27, 1971, p. 50.

62. Daniloff, *The Kremlin and the Cosmos*.

63. Theodore Shabad, "Russians Indicate Rocket Specialist Heads Space Effort," *New York Times*, July 14, 1974, p. 6.

64. "Akademik Vladimir Nikolayevich Chelomey," *Pravda*, December 12, 1984, p. 3.

65. These included the following major pieces: V. Rodikov, "General'nyy konstruktor" ["General Designer"] in *Zagadki zvezdnykh ostrovov: kniga chetvertaya* [*Mysteries of Starry Islands: Book Four*], ed., F. S. Alymov (Moscow: Molodaya gvardiya, 1987), pp. 53-100; and V. Rodikov, "Akademik Chelomey i ego vremya" ["Academician Chelomey and His Time"] in *Zagadki*

zvezdnykh ostrovov: kniga pyataya [*Mysteries of Starry Islands: Book Five*] ed., F. S. Alymov (Moscow: Molodaya gvardiya, 1989), pp. 4-36. Rodikov also wrote several other articles on Chelomey in 1985-1989 in magazines such as *Kryl'ya rodiny* (*Wings of the Motherland*), *Aviatsiya i kosmonavtika* (*Aviation and Cosmonautics*), and *Tekhnika-molodezhi* (*Technology for Youth*).

66. Sergei N. Khrushchev, *Nikita Khrushchev and the Creation of a Superpower* (University Park, Penn.: Pennsylvania State University Press, 2000). An earlier shorter memoir about his father had appeared in 1990: Sergei Khrushchev, *Khrushchev on Khrushchev: An Inside Account of the Man and His Era* (Boston: Little, Brown and Company, 1990).

67. Vladimir Polyachenko, "Chelomey and Korolev – Cooperation and Competition," *Spaceflight* 53 (July 2011): 271-277.

68. For my first stab at this history, see "The Almaz Space Station Complex: A History, 1964-1992." *Journal of the British Interplanetary Society* 54 (2001): 389-416 and 55 (2002): 35-67.

69. CIA, *Soviet Military Capabilities and Intentions in Space*, National Intelligence Estimate, NIE 11-1-80, original classification Top Secret, August 6, 1980, p. 20.

70. These books included: (co-authored with R. F. Appazov), *Ballistika upravliaye-mykh raket dal'nego deystviya* [*Ballistics of Guided Long-Range Missiles*] (Moscow: Nauka, 1966); *Vvedeniye v mashinoye proyektirovaniye letatel'nykh apparatov* [*Introduction to Machine Design of Flying Vehicles*] (Moscow, 1978); *Algoritmy diagnostiki teplovykh nagruzok letatel'nykh apparatov* [*Diagnostic Algorithms of Thermal Stresses on Flying Vehicles*] (Moscow: Mashinostroye-niye, 1983).

71. For the originals, see: M. Vasil'yev, ed., *Shagi k zvezdam* (Moscow: Molodaya gvardiya, 1972); M. P. Vasil'yev, *'Salyut' na orbite* (Moscow: Mashinostroye-niye, 1973).

72. M. Vasil'yev, "Sputnik: nachalo kosmicheskoy ery" ["Sputnik: Beginning of the Space Age"], *Pravda*, April 10, 1972. Other articles were published in *Izvestiya* (December 28, 1972) and *Krasnaya zvezda* (April 12, 1974).

73. Pierre Dumas, "Un Train Spatial Habite Vers Mars en 1978," *La Recherche Spatiale* 11 no. 3 (1972): 26.

74. S. Yu. Protsyuk, "Tekhnichna khronika khto keruye programmeoyu osvoyen-nya kosmosu v srsr?" ["Technical Chronicle: Who Runs the Programme of Mastering Space in the USSR?"], *Ukrainian Engineering News* 23 nos. 3-4 (1972): 60-72.

75. Albert Ducrocq and Martine Castello, *Le livre d'or de la science* (Paris: Solar, 1977).

76. Victor Yevsikov, *Re-entry Technology and the Soviet Space Programme: Some Personal Observations* (Falls Church, Va.: Delphic Associates, 1982).

77. C. Wachtel, "The Chief Designers of the Soviet Space Programme," *Journal of the British Interplanetary Society* 38 (1985): 561-563.

78. Others on the "editorial council" included Mstislav Keldysh, Boris Raush-enbakh, Georgiy Tyulin, Vladimir Barmin, Valentin Glushko, Nikolay Pilyugin, Konstantin Bushuyev, K. Kolesnikov, Yuriy Mozzhorin, Sergey Okhapkin,

Konstantin Feoktistov, A. Yeremenko, and V. Sokol'skiy. The chief editor-compiler was (later) famous space historian Georgiy Vetrov. See M. V. Keldysh, ed., *Tvorcheskoye naslediya akademika Sergeya Pavlovicha Koroleva; izbrannyye trudy i dokumenty* [*The Creative Legacy of Sergei Pavlovich Korolev: Selected Works and Documents*] (Moscow: Nauka, 1980).

79. See for example "Nashi interv'yu" ["Our Interview"], *Zemlya i vselennaya* no. 6 (1987): 2-5; Yevgeny Andrianov, "The First Sputnik," *New Times*, October 12, 1987, pp. 24-25.

80. A. Tarasov, "Polety vo snye i nauavu" [Flights in Dreams and Reality], *Pravda*, October 20, 1989, p. 4.

81. V. P. Mishin, "Pochemu my ne sletali na Lunu" [Why Didn't We Fly to the Moon?], *Znaniye: kosmonavtika, astronomiya* no. 12 (1990): 3-43.

82. *Russian Space History*, Sale 6516 (New York: Sotheby's, 1993); John Noble Wilford, "Space Papers Going on Sale," *New York Times*, December 5, 1993, p. 36.

83. These included: "Zenit – The First Soviet Photo-Reconnaissance Satellites," *Journal of the British Interplanetary Society* 50 (1997): 441-448; "ZENIT: Corona's Soviet Counterpart," in Robert A. McDonald, ed., *Corona: Between the Sun & The Earth: The First NRO Reconnaissance Eye in Space* (Bethesda, MD: The American Society for Photogrammetry and Remote Sensing, 1997), pp. 85-107; "The Soviet Response to CORONA," in Dwayne A. Day, John M. Logsdon, and Brian Latell, eds., *Eye in the Sky: The Story of the CORONA Spy Satellites* (Washington, D.C.: Smithsonian Institution Press, 1998), pp. 157-170; and "Black 'Amber': Russian Yantar-Class Optical Reconnaissance Satellites," *Journal of the British Interplanetary Society* 51 (1998): 309-320.

84. "Peter A. Gorin Obituary," http://www.legacy.com/obituaries/dailypress/obituary.aspx?page=lifestory&pid=123084975. (accessed May 31, 2012).

85. "Delegatsiya NASA v MAI" [Delegation from NASA at MAI], http://www.mai.ru/events/news/detail.php?ELEMENT_ID=28466. (accessed May 31, 2012).

86. Christoph Mick, *Forschen für Stalin: Deutsche Fachleute in der sowjetischen Rüstungsindustrie 1945-1958* [*Research for Stalin: German Experts in the Soviet Defense Industry, 1945-1958*] (Munich: R. Oldenbourg, 2000); Matthias Uhl, *Stalins V-2: Der Technologietransfer der deutschen Fernlenkwaffentechnik in die UdSSR und der Aufbau der sowjetischen Raketenindustrie 1945 bis 1959* [*Stalin's V-2: Technology Transfer of German Long-Range Missile Technology to the USSR and the Establishment of the Soviet Rocket Industry, 1945 to 1959*] (Bonn: Bernard & Graefe-Verlag, 2001).

87. Asif A. Siddiqi, "The Decision To Go to the Moon: The View From the Soviet Union," *Spaceflight* 40 (May 1998): 177-80 and 40 (June 1998): 227-30.

88. Asif A. Siddiqi, "A Secret Uncovered: The Soviet Decision to Land Cosmonauts on the Moon," *Spaceflight* 46 (May 2004): 205-13.

89. Yu. M. Baturin, ed., *Sovetskaya kosmicheskaya initsiativa v gosudarstvennykh dokumentakh, 1946-1964 gg.* [*The Soviet Space Initiative in Government Documents, 1946-1964*] (Moscow: RTSoft, 2008), pp. 269-273.

90. These have been published in 2004-2011 under the general title *Rockets and People* by the NASA History Office.

91. Rex Hall and David J. Shayler, *The Rocket Men: Vostok & Voskhod, The First Soviet Manned Spaceflights* (Chichester, UK: Springer-Praxis, 2001); James Harford, *Korolev: How One Man Masterminded the Soviet Drive to Beat America to the Moon* (New York: John Wiley & Sons, Inc., 1997).

92. Nataliya Koroleva, *Otets: kniga pervaya* [*Father: Book One*] (Moscow: Nauka, 2001).

93. Asif A. Siddiqi, "Cosmic Contradictions: Popular Enthusiasm and Secrecy in the Soviet Space Programme" in *Into the Cosmos: Space Exploration and Soviet Culture*, eds. James T. Andrews and Asif A. Siddiqi (Pittsburgh: University of Pittsburgh Press, 2011), pp. 47-76.

94. These documents are stored in RGAE, *fond* [fund] 4372, *opis'* [inventory] 81, *delo* [file] 1239, *listov* [pages] 44-48.

95. G. S. Vetrov and B. V. Raushenbakh, eds., *S. P. Korolev i ego delo: svet i teni v istorii kosmonavtiki: izbrannyye trudy i dokumenty* [*S. P. Korolev and His Affairs: Light and Shadow in the History of Cosmonautics: Selected Works and Documents*] (Moscow: Nauka, 1998)

96. M. V. Tarasenko, *Voennyye aspekty Sovetskoi kosmonavtiki* [*Military Aspects of Soviet Cosmonautics*] (Moscow: Nikol, 1992)

97. V. I. Ivkin and G. A. Sukhina, eds., *Zadacha osoboy gosudarstvennoy vazhnosti: iz istorii sozdaniya raketno-yadernogo oruzhiya i raketnykh voysk strategiches-kogo naznacheniya (1945-1959 gg.): sbornik dokumentov* [*A Goal of Special State Importance: From the History of the Creation of Rocket and Nuclear Armaments and the Strategic Rocket Forces (1945-1958)*] (Moscow: Rosspen, 2010)

10

Urban cosmonauts and space historians

by David J. Shayler

According to a dictionary, the above title can be defined as:

Urban – *living in a town or city*
Cosmonauts – *Soviet astronauts*
(Space) Historian – *a researcher or chronicler of past events (in the exploration of space)*

This could be interpreted to mean several former Russian space explorers who now live in residential areas and write about their experiences but, in fact, it describes an international group of enthusiastic, mostly amateur, writers living in large towns and cities across the globe who have spent years investigating the mysteries and gaps in information of a once secret space programme. This usually required countless hours of dedication, normally after concluding a 'normal' working day, and also involved a dogged effort to get the findings published.

MY SLEUTHING STORY

In my research I do not differentiate between the American, Russian, or Chinese programmes; I study the history and operations under the topic of human spaceflight from whichever country it originates. It just so happens that this information could vary considerably in different parts of the world. With the Soviet and now the Russian space programme the difficulty in obtaining detailed information has created, over a period of many years, a worldwide network of contacts and colleagues – many of whom have become very good friends. My involvement in this network of 'Soviet space sleuths' has always been a most rewarding experience. I am not a professional journalist, neither do I have a scientific background or formal historian qualifications. I remain a self-taught enthusiast who strives to do a professional job in my writing career.

10.1: David Shayler with some of his spaceflight files.

Information scarce at first

Prior to the early 1990s, trying to find out anything about Soviet space activities was no mean challenge. Even after the breakup of the Soviet Union and the creation of a 'new Russia', what emerged was still clouded in uncertainty.

During the late 1960s, when I took an interest in the activities and achievements of cosmonauts, it was difficult to discover even the briefest item of detail behind the headlines and carefully scripted news releases. To find a Russian book on the subject was very rare, and if you did it would be expensive. Back then Western books were sparse in their coverage of Soviet space activities. Throughout the 1970s and for most of the 1980s even the smallest research achievement was the result of many hours of painstaking research and thoughtful reflection, usually spread over months or perhaps even years. This could at times become tedious, with dead ends and misinterpretation, but slowly mysteries were solved. Scribbled notes were updated and filed, suggesting a new line of study. Letters were published or full articles appeared in a space-related magazine to make known the latest findings and to pose new questions.

Letters to the editor

James Oberg of the United States was one of the first researchers to pose questions or offer theories on the early Soviet manned space flights. Jim did not merely report the facts or repeat the official line, but advanced his own conclusions and theories. His letters and articles in the magazine *Flight International* and the British Interplanetary Society's (BIS) *Spaceflight* in the early 1970s were a catalyst for other writers to take a deeper look at the programme. The 'golden years' of space sleuthing were about to flourish, although it would be several years before book publishers took notice of this research.

It soon becomes apparent to any author that you do not become rich from writing about the space programme. But in studying the Soviets that was not the point; it was the satisfaction of announcing a finding that filled in a gap and provided others with a basis for further study. Whilst serving in H.M. Forces, I learnt that there is usually only one way to achieve an objective – and it is invariably the hard way. Space sleuthing most definitely falls into that category.

FIRST INFLUENCES

There can be a few key people in one's early life that provide subtle introductions to paths which develop in later years. I recall a teacher at secondary school in the late 1960s who was influential in developing my interest in space exploration, especially with regard to the Soviets. He was a history teacher who had studied and could teach Russian; albeit not in our school. It was his introduction to reports on the latest space shots in Soviet newspapers that inspired my interest in the cosmonauts.

Another key figure was the British Interplanetary Society's Kenneth Gatland, who wrote a book on manned spacecraft in 1967 [1]. I found a copy in my school library the following year. From that book, which became the first volume in my own space library, I learnt about the early years of the Space Age. The descriptions of pioneering Russian and American missions were accompanied by stunning full-colour artwork. To a thirteen-year-old boy eager to pursue a career as an engineering draughtsman, the technical illustrations were impressive.

Of my secondary school years, I have clear memories of the memorial service for the three Apollo 1 astronauts who died in the pad fire on 27 January 1967; and of the TV transmissions from Apollo 7 in October 1968, sitting in front of our small black and white set enthralled with the "Wally, Walt and Don show" from "high atop of everything". And when Apollo 8 flew in lunar orbit on Christmas Eve I was hooked for life. My interest in what was at that time referred to as 'manned spaceflight', soon became a hobby, then a passion, and eventually a career.

The early months of 1969 brought regular TV news items about future Apollo missions heading to the Moon. These were accompanied by a growing number of newspaper and magazine articles, special 'Moon-flight' supplements and books that recounted earlier NASA missions to provide context for the momentous events of that year. There were also brief and tantalising reports of cosmonauts and their achievements. I was intrigued. Could the Soviets be preparing to go to the Moon at the same time as the Americans? Most reporters seemed to think so.

Soviet response to Apollo?

There was the question of what precisely the Soviets were planning to do in response to Apollo. There was lots of speculation but very little solid information in the British media. Fortunately, my Russian speaking history teacher came to my aid and set me on course for many decades to come. As a result of my questions he brought in copies of newspapers such as *Pravda* and *Izvestia* for me to browse. I was unable to read the reports, but I could enjoy the pictures and the spell-binding nature of a paper from the enigmatic Soviet Union was magical. My fascination was heightened in January 1969 when Soyuz 4 and Soyuz 5 generated headlines in British newspapers by performing the first-ever docking of two manned spacecraft, and then a spacewalk by two of the Soyuz 5 cosmonauts to Soyuz 4 for their return to Earth. The spacecraft were said to have formed the first "space-station" although, as I realised later, this was stretching things a little. My teacher translated the information presented in the Russian papers, but it was evident that there was much more to the story than was being reported.

Growing interest in the Soviets

When the world's first space station, Salyut 1, was launched in the spring of 1971 I hoped to learn more about the Soviet space programme. The failure of the Soyuz 10 cosmonauts to enter Salyut and the deaths of the Soyuz 11 cosmonauts at the end of a record breaking three week stay in space did generate news reports, but not for the right reasons.

During these years I had left school, completed a basic engineering draughtsman course and joined H.M. Forces, so finding time to study space developments was, by the nature of things, difficult. But I still found time to keep up-to-date. By 1973 I had subscribed to *Spaceflight* magazine and became fascinated with articles relating to the Soviet space programme. By the mid-1970s I could at long last devote more time to my interest in human spaceflight. However, by then there was only one flight on the American manifest, the Apollo-Soyuz mission in 1975. Consequently, I became more interested in learning about what the Soviets were doing and how I could source the latest information.

I subscribed to *Soviet Weekly*, the *Soviet News* bulletin, and even got regular news releases from the Novosti News Agency. I also obtained the UK aerospace magazine *Flight International*, which had occasional reports on the latest Soviet developments. By visiting central libraries I could also read the American magazine *Aviation Week & Space Technology*. At last my collection of Soviet notes, cuttings and reports was expanding. I began to compile my own notes and records. My scrap books evolved into a useful archive and resource. Full-time employment still had to take priority, but every spare moment was devoted to my research. This balancing act would last over 25 years until I was finally able to devote my full attention to a space writing career.

An early role-model

In 1975 I began a regular communication with Gordon R. Hooper, who was writing a series of articles on the Salyut space stations and the cosmonaut team for *Spaceflight*. Gordon worked in the banking industry, so was writing as a hobby. He had joined the BIS in 1973 and started writing in 1974. His series 'Missions to Salyut' began with the Salyut 3 programme in 1974 and ran to the end of the first Salyut 6 expedition in March 1978. By recounting the daily activities of cosmonauts using the official news releases, his articles were a handy summary of each expedition. These were excellent inspirational accounts at a time when I was trying to find my own level and direction of writing. Gordon was a very helpful correspondent in those early years, and raised my interest in all things Soviet.

In addition to reporting on the Salyut missions, Gordon wrote biographies of the flown Soviet cosmonauts, covering their lives and careers in the period 1961-1977. In 1977 he privately published updated biographies for all flown cosmonauts, and even included summaries of the men dubbed 'the missing cosmonauts' – names which had been unofficially identified as potential cosmonauts. He used a duplicator system and produced the booklet as a stapled A5 document with a card cover. It was very basic, but effective, and it made me wonder what I might be able to achieve using my own archive.

Gordon was one of the first *Spaceflight* contributors to author a regular series on Soviet activities, popularising the field to other members. He was made an Associate Fellow of the BIS in May 1979 and nominated a Fellow in January 1984. After a four year period of additional research, he published *The Soviet Cosmonaut Team* in 1986 through his company GRH Publications. This was a significant update to his earlier booklet. Although the book was a ground-breaking, it was published as great changes were underway in the Soviet Union, and the appearance of new sources inside Russia led him to expand it into a two-volume work in 1990.

JOINING THE 'BIS'

Partly from Gordon's articles and the enjoyment of reading issues of *Spaceflight*, I joined the British Interplanetary Society in 1976. With the encouragement of the editor of *Spaceflight*, Kenneth Gatland, I began submitting correspondence and small space news reports to the magazine that year. To my pleasant surprise they were published! I then began working on my first fully-fledged article, choosing a Soviet theme which I had been researching for a year. This reported on the different radio identification call-signs used by cosmonauts during their missions. It was published in January 1977 [2]. Instead of naming each spacecraft, as the American astronaut tended to do for Apollo, Soviet cosmonauts had their own personal call-sign which was only used on a mission if they flew as commander. The other members of that crew adopted their commander's call-sign adding a number one or two as required. This system is still used by Soyuz crews flying to the International Space Station.

It was quite rewarding to see a named article in print and gratifying to contribute just a small piece to the historical jigsaw of Soviet cosmonautics. I will always be grateful to the BIS for supporting my work in their publications over the last 40 years. The rewards were not financial, but the enjoyment, experience and satisfaction were priceless.

A true Yorkshireman

By the late 1970s I was in regular correspondence with Neville Kidger who had taken over from Gordon Hooper in reporting on the Salyut missions in *Spaceflight*. Born in 1953, Neville was very much a true Yorkshireman. He joined the BIS in 1978 and his dedication over the next three decades wrapped up the missions to Salyut 6, covered all the missions to Salyut 7 and Mir, and the first 21 expeditions to the International Space Station.

10.2: Neville Kidger with NASA astronaut Mark Kelly.

Determined not just to report the basic facts from Novosti press releases or Radio Moscow broadcasts, Neville sought to reveal the deeper story behind the news. By broadening each instalment of the series, he gave fresh insights into the operational aspects and infrastructure of long duration spaceflight operations. His very accurate and informative monthly space station reports were even more remarkable when one recalls that he was in full-time employment and devoted significant time to his other passions in life – his family and Leeds United. In December 2003, in recognition of his outstanding commitment to the society, Neville was awarded the BIS Sir Patrick Moore medal for his reliable and well-written 'Space Station Reports' over what was then a period of 25 years [3].

News from Moscow

During the mid-1970s, I regularly listened to the evening news broadcast from Radio Moscow. The reporting of the latest developments in space operations and technology by their correspondents, especially Boris Belitsky, added to the mysteries of the space programme. I put my ear to the speaker to listen through the static and 'read between the lines' to learn what the latest mission had accomplished. Fond memories of sitting in the kitchen night after night during 1976 and 1977 listening to reports about Soyuz 21, 22, 23 and 24 remain with me. It was from Radio Moscow that I first learnt of the decision to fly 'guest cosmonauts' from Eastern Bloc countries. Seeking to obtain details on the Soviets indirectly, I sent letters to the London embassies of Bulgaria, Czechoslovakia, Cuba, East Germany, Hungary, Mongolia, Poland, Romania and, a little later, Vietnam. As replies arrived between 1978 and 1981 I was very surprised. Some included only a single press release or solitary black and white picture, whilst others were a wealth of information and booklets which were welcome additions to my collection.

A new contact

Between 1977 and 1980 I had authored (or co-authored) a series of articles with BIS members Curtis Peebles, Phillip Snowdon and Andrew Wilson about the American astronaut team. Over the years, I had seen letters to the correspondence section of *Spaceflight* which had generated informative replies and useful contacts. I submitted several letters myself and received dozens of responses, including a fascinating and neatly handwritten one from Rex Hall in London. That initial letter from Rex in May 1980 was the starting point of regular contact between us over the next 30 years that would include working with the BIS; various privately published documents; a range of cooperative projects with other authors across the world; three books on the Soviet programme; talks and symposia; supporting a number of cosmonaut visits to the UK; and a memorable trip to Russia. That letter was one of those rare moments in life that alter destiny. It was the support, encouragement, enthusiasm and calm personality of Rex that guided me deeper into the world of cosmonautics and gave me confidence to develop my own contributions to Soviet space sleuthing.

10.3: Renowned cosmonaut expert Rex Hall.

THE MAKING OF A NON-COSMONAUT

It is a paradox that those who thought they knew Rex, were surprised to find out how much they didn't know about him. That was simply the character of the man, always generous and helpful to someone who sought his assistance but never pushy or self-promoting. He was never one to reveal much about himself – opting instead to focus more on the task at hand. Rex was widely recognised as the world's leading authority on the Soviet cosmonaut team. Indeed, it has often been acknowledged that he knew more about the cosmonauts than the cosmonauts themselves!

Born in London in 1946, Rex graduated from the country's first comprehensive school in Holland Park. Whilst in school, Rex became caught up in what was widely being called the Space Race. In the summer of 1961 he visited a Soviet exhibition in London that displayed a model of the Sputnik 3 satellite that was incorrectly labelled as a Vostok manned spacecraft. With a desire to obtain more in-depth information he sought biographical details on the flown cosmonauts, writing to the Soviet Embassy in London but never receiving a reply. In a bookshop a few years later he discovered a set of cards on Soviet space achievements and renewed his efforts to find out more.

From 1965 Rex worked for an import and export company in London for eighteen months, then spent the next eighteen months in the United States as an employee of Chrysler relocating their vehicles across the country. He followed this by a period of eighteen months backpacking in India, Afghanistan and other places thereabouts. He accepted a wide range of jobs to fund his adventures, including at one point working as a marble fountain cleaner for a Maharaja. Returning to the United Kingdom in the early 1970s he worked for a company in the stock exchange for several months prior to becoming a youth worker in London. It was in this field that he would develop his career over the next four decades, focusing initially on youth unemployment around the Tower Hamlets area of London. He later developed a national UK government-funded programme supporting youth initiatives using sport facilities and resources.

A dedicated student

Joining the BIS in 1974, Rex had already begun to collate his own researches in the form of a very effective filing system, and over the years he built up a unique (and much envied) library on the Russian space programme. His first visit to the Soviet Union was to a freezing Moscow in February 1971 as part of his youth educational work. Over the next few years his interest in the Soviet programme grew – mainly because it was so secret and difficult to find anything about it. This challenge suited Rex's character perfectly; his determination to unearth as much as he could would occupy his limited spare time for the next four decades.

From his own investigations Rex compiled the basic facts on each member of the cosmonaut team, including a list of 'missing' backup members; he also realised that other researchers were doing the same. From this early research he established a network of contacts and correspondents across the world. Over time his meticulous research made him a leading expert on the Soviet cosmonaut team.

His ongoing support of the BIS saw Rex elected as a Fellow in 1986; become a member of the Council in 1995; and serve as President of the Society between 2003 and 2006. It was during his term as Society President that Rex was instrumental in arranging for the 2008 International Astronautical Congress to be held in Glasgow.

It is surprising that with so much involvement with the BIS and his investigations of cosmonauts Rex continued to work full-time in youth education. A key factor in his success and achievements was the support of his wife Lynn.

"That book"

As the years rolled by, if there was a book or booklet published on Soviet manned space exploration, Rex would have a copy. All except one – and that book was in my collection. Back in 1980 I wrote to the embassy in London asking for information on the upcoming flight of a Vietnamese cosmonaut to Salyut 6. Never really expecting a reply, I was amazed at what arrived in the post a short time later – a small pre-flight propaganda booklet on the selection and training of the two Vietnamese cosmonauts. It was poorly printed, with blurred photos and written in Vietnamese.

At the time I did not realise the unique nature of this publication and filed it with other documents from the East European, Cuban and Mongolian embassies. A few years later, when I showed Rex the book, he was really eager to obtain a copy for his own collection. He wrote to the Vietnamese embassy, but much to his frustration nothing came back. In the thirty years since receiving my booklet we never saw or heard of another copy. It served as an excellent potential bartering item for many years, which Rex took in his usual humour. This little booklet became known between us as "that book" and it still has pride of place in my collection.

THE SOVIET FORUM

One of the most rewarding aspects of being a member of the BIS has been attending its Soviet Forum as often as I could since the early 1980s. It was my participation in these technical meetings that provided contact with a number of leading Soviet space historians and greatly enriched my own research studies.

In 1979 the late Anthony Kenden, a pioneering investigator of military space activities, persuaded the society to hold an informal meeting of members interested

10.4: Informality prevailed at the 2012 BIS Soviet Forum.

in discussing the latest developments under the title 'Technical Forum'. The inaugural meeting was held in January 1980 with the topic of 'The Soviet Space Programme'. Some of those in attendance were Phillip Clark, Rex Hall and Nicholas Johnson; the latter flying over especially from the United States.

The meeting was an instant success and soon became known as the annual 'Soviet Technical Forum' – normally held on the first Saturday of June. These meeting were always relaxed, and new speakers were encouraged to present a paper, no matter how brief or informal. The Forum is an opportunity to meet old friends and establish new ones, chat about recent developments, learn new revelations, and recall past events as well as future plans. Rex played a key role in its organisation as chairman each year. Although many of the audience attended several of the Forums over the years, Rex was the only one to attend all thirty from 1980 through 2009.

'Clarkisms'

Phillip Clark was another regular, and his attention to detail in his research was legendary. Each year Phil would present his paper on some obscure aspects of the Soviet (or Chinese) programme. With a degree in mathematics and computing, his

10.5: Space sleuth Phil Clark in 2006.

careful analysis and statistical evaluation could 'fill in gaps' in official information, and over the years his reports became an important element of Soviet space sleuthing.

For each presentation, Rex would make a short introduction and then remain out of sight of the audience at one side of the meeting room, but in view of the presenter to ensure that the meeting kept to the timetable. When Phil displayed his 'equation slides' sometimes groans could be heard from off-stage, suggesting that Rex was seriously contemplating having a lie down to recover before introducing the next speaker. Friendly banter was part and parcel of the Technical Forums – especially with regular presenters who became known as "the usual suspects". Though very appreciative of Phil's contributions, Rex's sense of humour was always quietly in evidence. After Phil appeared on a BBC documentary on the Soviet programme, a photo of Phil inside the Kaliningrad control centre in Russia was shown at the next Forum and Rex dubbed him one of the "missing cosmonauts". At the 1981 Forum Phil had suggested that the Soviets were possibly about to end their series of Progress resupply craft. However they did not, and in 1989 as a new freighter left Earth for Mir a plan was hatched between us. We knew Phil was going to deliver a talk in the Midlands in August of that year. As Rex would be unavailable, and I was going to be in the United States, we asked a member of my family who was not so well known to Phil to attend and ask him when he thought the Soviets would stop flying Progress resupply craft. Phil caught on to the joke immediately and asked how I was getting on in America!

Another fond memory of these annual BIS meetings was the gathering of some of the speakers at Rex's home to talk about points raised during the day. This informal social gathering was an excellent place to discuss some of the lasting mysteries of the programme.

Soviets in Britain

Shortly after Gagarin's historic spaceflight Rex had been fortunate to visit a Soviet space exhibition in the summer of 1961. It was around the same time as the world's first spaceman made his short trip to London and Manchester. Almost 30 years later I took the opportunity to tour a Soviet exhibition on space exploration hosted in Birmingham. A variety of spacecraft displays ranging from full-sized mockups to scale models and memorabilia was a great opportunity to examine Soviet space hardware up close for the first time and in my home town.

I had recently returned from the United States where I toured the training facilities at the NASA Johnson Space Center and conducted research at its History Office and at nearby Rice University in Houston. I had also undertaken interviews with several astronauts, so it was a veritable bonus to get the opportunity to have my first meeting with a Soviet cosmonaut. Engineer cosmonaut Viktor Savinykh was touring with the UK exhibition, along with liaison officer Valery Terekhin from the Russian embassy in London. Terekhin was my contact in arranging an interview with Savinykh at his hotel the evening before I visited the exhibition. In fact, this was the first of a number of meetings with Terekhin over the next couple of years, who I later found out was a serving Russian Air Force officer – and apparently also a member of the KGB! The interview and meeting went very well, and Savinykh seemed surprised and delighted that I had a copy of a Russian book on Salyut 6 for him to sign.

The next day I arrived early at the exhibition, along with several other local space enthusiasts including Andy Salmon. I had arranged a second interview with Savinykh prior to touring the displays. As it was so early, the venue was sparsely populated. I asked Terekhin if, since I was researching a book on spacesuits, I could examine the display of a Russian Orlan EVA suit in more detail; perhaps even opening some of the thermal covering to reveal the metal hard upper torso beneath. He said that would be fine, so I went ahead and took some great close up images. Whilst walking around the exhibition we noticed another person following us, seeming at first to be another visitor, but we had our doubts as he seemed to be pointing his camera at us instead of the exhibits. When I asked Valery about this, he nodded and quietly advised us not to be concerned – pointing out with a broad smile "he is merely updating your files for Moscow". This news did little to settle our concerns at the time though. As a Soviet space watcher I was aware of the interest that this attracted, especially after Rex had joked on more than one occasion that we all must have expanding files in the *FBI* by now.

A few months later, Rex and I were in London at Earls Court attending another exhibition at which American astronaut Walter Cunningham and Soviet cosmonaut Yuri Romanenko were appearing. Once again Valery Terekhin was contacted and arranged for Rex and myself to interview them both. It was a fascinating meeting, seeing two space explorers from different countries with a common experience of seeing our planet from orbit.

The interview with Romanenko was memorable for the strange protocol that we had to observe. We knew he spoke excellent English, since he had been assigned as a support cosmonaut on the Apollo-Soyuz Test Project over a decade earlier. He knew we were aware that he could speak English (as we heard him talking to Cunningham a short time before the interview) but we had to ask questions and receive his replies via an interpreter. This was fine until the interpreter had difficulty in understanding our questions on EVA. At this point Romanenko said in impeccable English "Extra Vehicular Activity means taking a walk in space", and then finished his response in Russian. Rex and I later laughed at the silliness of it all.

It is fascinating to get information first-hand, but you must always be alert to the distinction between factual recollections and what might be termed 'space stories'. A few years later I interviewed Georgi Grechko in Manchester. He spoke in excellent English about his 1977 EVA, saying how Yuri Romanenko had wanted to take a look outside too, and when Romanenko floated to the hatch to poke his head and shoulders out Grechko observed that his colleague wasn't tethered and grabbed him to save him from floating away. At our earlier meeting, Romanenko had made no mention of this. In October 2007 I had the opportunity to speak with Romanenko again and so I told him of Grechko's recollection of the events on the EVA. Romanenko replied that he did not remember it the same way. He was adamant that he was attached by a safety tether the whole time but added, "Georgi always liked to exaggerate a good story!"

'Astro Info Service' is created

By the early 1980s, after several years of writing articles for the BIS, I was interested in publishing my own documents and reports. I recalled Gordon Hooper's privately produced cosmonaut book in 1977, and in 1982 I asked a high-street printer near to home about self-publishing at a reasonable cost. I had developed an idea for a regular 'fanzine' type of periodical that would cover the development, news and operations of human spaceflight. At the same time I had compiled a booklet based on my notes and biographical data summarising the 1963 NASA Group 3 astronaut selection, the first of a planned series. These were too large for magazine articles and yet too small and specialised for a book publisher, therefore private publication seemed the best option. I also decided to set up a company in order to focus on my writing efforts.

Whilst remaining in full-time employment, I established the Astronaut Information Service (AIS), almost immediately shortened to Astro Info Service. Initially intended to record the latest news, assignments and careers of American astronauts, the scope soon expanded to include European astronauts and Soviet cosmonauts. AIS started trading on 4 October 1982, on the 25th anniversary of the launch of Sputnik. Its aim was (and remains) to provide accurate information, research and publications on the human exploration of space.

A point in the heavens

For some time Rex Hall had been assisting in the updating of my own archive of cosmonaut data. He readily agreed to cooperate on compiling a small booklet on the cosmonaut team based on his meticulous research and we planned for publication to mark the 25th anniversary of the creation of the original team in 1960. The booklet included a collection of selection details and biographies complemented by a list of experience, together with a log of Soviet spaceflights to date. It was published as *The Soviet Cosmonaut Detachment 1960-1985*. This was updated at least five times over the next 20 years. The design of each entry for a cosmonaut was based upon Rex's original and highly effective note form recording system, which presented the basic facts and information for each person.

By early 1985 I had decided that a regular publication on the Soviet programme could complement the forthcoming cosmonaut book. A 'fanzine' type of periodical seemed the way forward. Its name took a little thought. When I compiled cosmonaut Alexei Gubarev's entry for the cosmonaut book I noted his call sign 'Zenit' (Zenith) and figured that it looked and sounded very 'Soviet'. A dictionary gave the meaning as a "point in the heavens". It seemed to fit, and so *Zenit* was born.

I discussed the idea with Neville Kidger for a while. He was enthusiastic about a new contemporary magazine on the Soviets, and agreed to be its editor and primary compiler. I would add articles, source other authors, and organise the production and distribution. Being in the days prior to computer-based desk-top publishing, each A5 issue was typed, photocopied and stapled. The first issue was published in June 1985 and was produced bi-monthly until December 1988 (#22). From January

ISSUE No.28
JUNE
1989

ZENIT

EDITOR:-
N. Kidger

SUBSCRIPTIONS:-
D.J. Shayler,
Astro Info Service,

10.6: The redesigned cover of *Zenit*.

1989 (#23) until December 1991 (#58) it was a monthly production owing to its popularity. The production ended with #64 in December 1992. We had reverted to a bi-monthly rate for the final year due to uncertainly in the new Russian space programme.

One of the objectives of *Zenit* was to use contributions written by leading Soviet space watchers to inspire new authors to contribute. Analytical reviews by Rex Hall (cosmonaut updates), Philip Clark (satellites, launch vehicles) and Andy Salmon (science articles and planetary exploration articles) alongside various contributions from myself and in-depth articles by Neville rounded out each issue; it was certainly a team effort.

DIGGING IN THE DUST

It is surprising how research in one area can suddenly lead in another direction. In 1988 I made my first of what would prove to be over a dozen research visits to the Johnson Space Center near Houston, Texas. For three weeks I was able to conduct interviews and personal research, as well as enjoy guided tours of the site. One of the most visited places was the JSC History Office Archive in Building 420, in one of the more inconspicuous locations at the back of the complex. It stored documents from older programmes and retired employees – in some cases the documents had literally been saved from the recycling bin.

Within the archives of the JSC History Office was a small collection of files about the Soviet space programme. They included a collection of papers, reports (including NASA technical translations) and small documents on various aspects of the Russian programme. It was in this collection that I found three photos which set me on a research journey into Soviet space/aviation history that has continued for almost 25 years. The images depicted what appeared to be a spherical capsule under preparation. At first glance it resembled a Vostok 'sharik' capsule with the letters CCCP painted on the side, but on the back was written: "USSR balloon flights 1933 ascended 11.8 miles into the stratosphere" and "USSR balloon flight 62,300 feet 1933". I recalled that in 1962 a parachutist (Pyotr Dolgov) had died in ejecting from a stratospheric balloon named Volga, and that prior to becoming a cosmonaut in 1966 Vasily Lazarev had participated in long duration Volga flights.

I photocopied the images and began to investigate the story more deeply at home. At the central library in Birmingham I obtained books on early stratospheric balloons, where I found brief details of Soviet record attempts in the 1930s. From these dates I scanned old copies of the *Telegraph* and *Times* newspapers, as well as the magazines *Aeroplane* and *Flight* for any contemporary news reports. It took a while but slowly the jigsaw came together. A few months later Rex showed me the latest additions to his impressive collection of Russian stamps and first day covers – including several commemorative stamps connected to the loss of a stratospheric balloon crew in the 1930s. In letters to colleagues around the world, I sought other information on these balloons. Significant assistance was offered from Vladim Molchanov in Russia, Bart Hendrickx in Belgium, and Colonel V. Tolkov of the Air Force Museum at Monino near Moscow.

Research results in a lecture

All this research work resulted in a paper presented at the Soviet Forum in the early 1990s [4]. Since then this research has continued, and further evidence has come to light including short film clips and more details from the Soviet files. I have reached the conclusion that the work carried out in the 1930s by the Soviets in the fields of pressure garments and life support systems, balloon flights into the stratosphere, the study of the upper atmosphere and the composition and effects of cosmic rays, and the later high altitude parachute descents, were as crucial to the creation of a Soviet space programme as the theoretical works of Tsiolkovsky and the rocket

10.7: Finally coming face-to-face with the Volga balloon.

experiments by Korolev. Through the stratospheric balloon flights of the 1930s the Soviets gained useful experience in ground support, hardware manufacture and testing, launch and recovery, tracking, public affairs, crew selection and training, scientific experiment selection and preparation, as well as flight operations and failure analysis. All these would have been very useful as the Soviets started to set up a programme to launch hardware and people into space three decades later.

My work came full-circle during June 2003 when I had the privilege to make my first visit to Russia. Part of the trip was to tour the Monino Air Force Museum where, in one of the displays, side by side, were the USSR and Volga stratospheric balloon gondolas. It was a special moment to view and to touch the actual hardware fifteen years after I first saw those three black and white images in a small file deep in the history archives almost hidden away in at the back of the JSC complex over in America.

Project Juno – a chance to fly?

In 1989 the Soviets invited the United Kingdom to participate in a mission to Mir in 1991. A newspaper advert calling for potential cosmonaut candidates specified that "no previous space experience was necessary", which was helpful to people like me. By the end of the 1980s the prospect of a British citizen flying in space any time soon was doubtful to say the least. Therefore when the call for Project Juno came with the proviso that "anyone could apply", it did not take long for me to decide I would take that leap of faith and apply. I sent off a letter of application. On receiving a package of application forms I eagerly filled them in and (after copying them for posterity) posted them off. From subsequent phone interviews I learnt that many of the 13,000 initial applications were rejected immediately with only a small percentage being sent the forms, so I was pleased to have survived the first hurdle. However, my progress was soon blocked by the fact that I did not have the desired career credentials.

Looking back over 20 years, I feel pleased that I attempted to become a British cosmonaut in what has proved to be one of the very few opportunities for someone from this country to apply for spaceflight training. I have used this experience as a 'cosmonaut applicant' (but not a candidate) in my educational work; inspiring youth to take the opportunity to achieve their dreams whenever they get the chance. It is, I explain, far better to try, than not to try at all and then look back years later and wish for what might have been.

Several years later I was able to meet all three members of the Soyuz-TM 12 Juno mission. In December 1998 I spoke briefly with Sergei Krikalev at the homecoming event for STS-88 at Ellington Air Force Base near Houston. I also meet with Helen Sharman on a few occasions. The first time was in late 1991 at a school in the West Midlands when I accompanied astronaut Bill Thornton on a lecture tour around the UK shortly after Sharman's flight. I met her again in April 2009 and enjoyed talking with her about her experiences on Mir. The commander of the mission was Anatoly Artsebarsky. We had several conversations during Autographica 9, in March 2007, about his career, flying in space and life on Mir. I mentioned in passing that I had applied to fly on Juno and that had I been successful he could have been *my* commander, which he thought was quite amusing, prompting me to wonder what he knew that I didn't!

I still treasure my UK Juno cosmonaut application forms and letter thanking me for my application; they are stored in pride of place alongside my Juno sweat-shirt emblazoned with the flight logo and, on the back, the slogan "I Applied to Ride".

Although I did not get to ride a Soyuz to Mir in 1991, I have been able to achieve another of my long held dreams: to see my name in print on a 'real' book in a high-street shop. Even better, in the space of just two years I had not one but three books published. In addition, I contributed to an American book containing biographies of the world's space explorers, which became a standard collectable volume for space enthusiasts. Those three books by leading publishing houses focused on the American space programme [5]. I was also very fortunate to work with Michael Cassutt in the United States (along with Rex Hall and Bert Vis) on all three editions of *Who's Who in Space* (1987, 1992 and 1999).

Meeting an unflown cosmonaut

It has always been a pleasure meeting in person any space explorer. The highlights of my research over the years have been talking to space pioneers from the early 1960s, the moonwalkers and various astronauts and cosmonauts. One of these memorable encounters featured an unflown cosmonaut whose identity was kept hidden in the days of the Soviet Union.

On 12 December 1992 I set up a trade stand at a space and astronomy exhibition in Sutton Coldfield Birmingham. It gave us an opportunity to promote AIS, its latest publications such as *Zenit* and the collection of cosmonaut biographies. The bonus at this show was the appearance of former 1965 military engineer cosmonaut Mikhail Lisun, who was expected to bring a Sokol pressure garment with him. I arranged with the organisers for an opportunity to talk with Lisun and looked forward to examining in close detail the Sokol suit used on Soyuz for ascent and entry. When Lisun arrived he did indeed have a Sokol suit with him, and allowed members of the public to put it on – though not to close the visor, which would have been dangerous since there was no life support system attached! This was too good an opportunity to miss. It was a tight fit because I had not been measured and had 'street clothes' underneath, which restricted the limbs as well as forcing a stooped stance due in part to its design. The attached gloves were really uncomfortable too. But donning the suit was a great experience and gave me useful knowledge for subsequent educational work and writing projects.

Later in the day I had the opportunity to talk with Lisun via an interpreter, and it was remarkable how frank he was about his career and experiences as a cosmonaut. I showed him the biographical sheet that I had compiled on him, developed from notes supplied by Rex Hall and snippets that Bert Vis had obtained four years earlier from the 'Star Town Information Group'. Fascinated that I wanted to know so much about him even though he had not flown in space, Lisun kindly went through the biography and suggested corrections and added a few more details about his

10.8: A rare opportunity to try on a Russian spacesuit!

career. It was quite remarkable that a former military cosmonaut should be so willing to talk about his previously secret assignments – although he left out some details and skirted around certain questions. Clearly there was still much to discover, but this was a great start and very satisfying in enabling us to update his biography in our records.

An expanding bookshelf

Clive Horwood began his career in the publishing industry in 1966 working for D. Van Nostrand Ltd, of which his father was then managing director. Ellis Horwood was made redundant and gave his own name to the successful scientific and technical publishing house that he founded in the early 1970s. Clive joined his father in 1976 and worked with him for 13 years until the company was sold in 1989. For the next five years Clive continued as managing director of Ellis Horwood Ltd until it closed in 1994. Shortly afterwards Clive created Praxis Publishing Ltd (*Praxis* – from the Greek meaning the practice of an art or skill) and in the next five years co-published over forty titles with John Wiley & Sons. This list included a number of books on the

Soviet space programme. Prior to that, in 1988, Brian Harvey's *Race into Space* was published by Ellis Horwood Ltd in its Space Science and Technology series and distributed by John Wiley & Sons. In 1996 Brian's book was updated and published by Praxis as *The New Russian Space Programme*. This was followed by other titles such as *The Mir Space Station* by David Harland.

In 1999 Praxis began co-publishing with Springer-Verlag, and has issued more than 300 titles, more than a third of which are directly related to space science and exploration. Overall, the company has produced an impressive range of authoritative titles on various aspects of human and robotic spaceflight, with a number of books reflecting Soviet and/or Russian space activities. Considering how little was available in the West not so many years ago, this series offers those interested in discovering more about the Soviet and Russian programmes the opportunity to access a reliable platform from which more in-depth research can be pursued.

Springer-Praxis are very proud of their range of space exploration books and believe the popularity derives from the breadth of a now-global space programme that is not just limited to the two former superpowers of American and Russia. Regarding the range of dedicated Soviet-Russian books in their list, Clive Horwood says that the popularity of the publishing list is due to "the quality of the authorship with specialist knowledge of the Russian space programme. They have provided the gravitas that the series needed from this perspective. The readership is indebted to them."

10.09: Clive Horwood (seated, with wine glass) with some of his authors in 2008.

It is also thanks to Praxis that a suitable outlet was provided for "authors with specialist knowledge" to see their sometimes decades of work finally in print.

TAKING THE PLUNGE

With several publishing contract projects confirmed, and a growing diary of space outreach bookings, I took the plunge and went full-time with AIS in 1999 and can honestly say that I have never looked back. It has been a challenging but rewarding roller-coaster ride. Spaceflight journalist Tim Furness once advised me that if ever I felt that writing had become a chore of meeting deadlines and paying bills, I should recognise that I had lost my edge. Thus far, for me, writing about spaceflight is still rewarding and certainly not a chore. With the advent of the Internet, 'digging in the dust' has become 'surfing the web'. However, I still prefer, whenever possible, sifting original papers and documents in search of that elusive morsel of detail. I also enjoy talking to participants, and direct contact via the web is certainly quicker and easier than writing lots of letters via 'snail mail'.

Trilogy of books

Some of the most rewarding experiences in recent years have been the opportunities to work closely with Rex Hall and Bert Vis on three major Soviet-Russian books for Praxis [6]. Knowing Rex's preference for remaining in the background, it was with some trepidation that I asked him in 1999 if we might compile a book together for Praxis. Surprisingly, he soon warmed to the idea and we finally agreed on a book to commemorate the early Soviet space triumphs up to about 1966. Our working titles for the book were 'The First Cosmonauts' or 'First in Space/Orbit' since those years saw the first satellite, first space rocket, first living creature in orbit, first man, first woman in space, first group flight, first crew, first civilians and first spacewalk.

Rex provided a lot of the early source material and whilst I compiled the draft text, he handled the biographies of key figures and sourced photographs. Combining his extensive archives of the early 1960s with my own contemporary resources was quite effective, and we worked well together. We had considerable help from fellow 'space sleuths' around the world. William Barry, who had studied at Oxford University in England and written his thesis on *Soviet Manned Space Policy 1953-1970*, offered a significant amount of support. A foreword for the book was agreed with Vostok 2 cosmonaut Gherman Titov but tragically he died just a few months before he was to supply the text. Fortunately, Titov's fellow 1960 cosmonaut Boris Volynov agreed to fill that gap.

The last volume together

At events, Rex would introduce me as the "Jackie Collins of the space book world" since I had authored so many titles. But I won the next round when the Soyuz book was published and we started work on the one about the Cosmonaut Training

Centre. Because Rex was some years older than me, I began to describe him as the "Barbara Cartland of the space book world"!

After the success of *Rocket Men*, I let Rex recover for a while, but not too long. I suggested a follow up book about the Soyuz would be a good idea as there was very little on the long-serving spacecraft available in the West. Research was simplified in one respect, because it was still operational and familiar to American astronauts on Mir and ISS programmes. In addition, from research trips to the United States I had copies of Apollo-Soyuz documents from Rice University that helped to reveal some interesting background information.

Another key contact for this Soyuz research was Bart Hendrickx in Belgium, who supplied a wealth of information as well as guidance as the text developed. As with *Rocket Men*, our *Soyuz* book was a pure joy to write though it did stress both Rex and myself at times to obtain that little extra item of elusive information we felt the book needed – or find a specific photo which just *had* to go in the book. Once again Rex's impressive network of contacts and friendship with many former cosmonauts helped.

Writing *Soyuz* was a rewarding challenge, as nothing like it had been published before. Several spaceflight participants later told Rex that they read our book during their preparations for a flight to the ISS, and the fact that it was later translated into Chinese greatly humoured him.

The third volume in the cooperative Rex Hall Trilogy, *Russia's Cosmonauts*, was the book describing the Yuri Gagarin Training Centre (TsPK) and cosmonaut training programme. It was co-authored with Bert Vis from the Netherlands. For several years Rex and Bert attended Association of Space Explorer congresses and they travelled to Russia to tour TsPK and other facilities in order to interview cosmonauts. As a result, they had assembled a wealth of information on the cosmonaut groups, their training programmes and facilities. Rex also had an extensive archive of information on the cosmonaut team itself. Bert was an authority on the various 'guest flights' on Russian Soyuz missions and on the Buran cosmonaut team. My focus was the EVA operations, flights by Russians on the Shuttle, and the visits by American astronauts to Mir. By combining these areas of interest, the layout of the book was easy to define. Convincing Rex to embark on a third venture was eased by his love for TsPK, and with both Bert and myself he was pleased to write on the infrastructure behind the crews and missions, and those who worked behind the scenes sometime for decades without recognition. Bert had an extensive gallery of photos taken at TsPK over the years identifying building and facilities, and how the facilities had evolved.

A trip to Russia

After several visits to the Cosmonaut Training Centre near Moscow, named for Yuri A. Gagarin, Rex and Bert invited me to join them for their June 2003 visit as part of our research for the book. This was my first visit to Russia and a packed programme ensured it was a truly memorable week.

The day before the trip began, I travelled to London to stay at Rex's overnight.

10.10: Star City's famous Yuri Gagarin monument.

He and I flew over on Saturday 14 June, were met at Moscow airport and driven to Star City where we were booked into the new Hotel Orbita for the week. A few minutes after our arrival we meet up with Bert Vis, who would join us for most of the week before he returned home. He had been in Russia for a few days interviewing former Buran cosmonauts before arriving at TsPK. During an evening meal at the cafeteria we received the week's schedule with Yelena Yesina, which looked quite impressive. Bert and Rex had been there many times before but for me it was my first trip and I was looking forward to anything and everything.

During the week official visits were made to the mission control centre (TsUP), Energiya (formerly OKB-1), the IMBP medical facility, the pressure suit company Zvezda, and the Monino Air Force Museum as well as a tour of the facilities (TsPK) at Star City. With Bert, I also visited the space memorial museum in Moscow under the Tsiolkovsky monument and completed two sombre but moving and memorable visits to the Ostankinskoye cemetery in Moscow and the cemetery located near to Star City.

There was so much to see and such a lot to take in. I had waited for over 30 years for this experience and was not disappointed. Since we were working on cosmonaut training, most of the week's visit focused on this research. Other visits and activities were invaluable for my own research on other projects such as spacesuits, EVA and stratospheric balloons. And, of course, I was eager simply to soak up and enjoy the whole experience.

Celebrating Tereshkova's flight

One of the more memorable events occurred on the evening of 16 June by attending the official celebrations for the 40th anniversary of the Vostok 6 mission on which Valentina Tereshkova became the first woman in space, where we were introduced to a large number of cosmonauts. After years of trying to research the cosmonaut team and match names and faces, here I was amidst a crowd of them, with Rex and Bert identifying them for me. At the end of the presentations, all five female cosmonauts from the 1962 selection were on the stage, which Bert later pointed out was the only selection group in Russia or America prior to 1978 for which all the members were still alive!

The earlier and very moving visits to the two cemeteries on Sunday 14 June were highlighted by seeing the beautifully crafted memorials and monuments to deceased cosmonauts. It was very sad to see so many famous names of Soviet space history, as well as some of the unflown cosmonauts. It was an honour to be shown some of the graves in the Star City cemetery by unflown cosmonaut Pyotr Kolodin, who seemed very pleased that foreigners knew of his deceased colleagues. Later at Lev Dyomin's grave we held back because members of his family were there, but when they saw us they invited us to join them and pay our respects to their famous family member, touched that we knew of his life and achievements.

Visits to Zvezda and the training facilities at Star City were other high points for me, especially squeezing into the rear hatch of an Orlan EVA suit. The stratospheric balloon gondolas USSR and Volga at Monino were (as stated earlier) a wonderful

10.11: Rex Hall having tea with cosmonauts Tereshkova and Bykovsky.

sight after so many years of research. But one of the most memorable days was the time Rex and I spent at the former Korolev OKB-1 design bureau, now known as Energiya. Unfortunately it was after Bert had left for home. On 19 June Rex and I were accompanied by Yelena Yesina on a formal visit to the Energiya facility. As we arrived we were met by Yuri Usachev, who accompanied the three of us and our official tour guides, a photographer and security around the famous facility. What was most memorable was the Soyuz and Progress production lines, where at least four spacecraft were in different stages of production. We could not touch, but we were allowed to insert our heads into one of the Progress shells briefly. What was truly remarkable was that in only a few months this vehicle would be docked to the ISS.

During the tour, Usachev was amazed at the places we were shown, remarking to Rex that he had worked at Energiya for years and had not been in some of the rooms we were taken into. In reply Rex, with his typically knowing look, just shrugged his shoulders and smiled broadly.

LOSING GOOD FRIENDS

Over the past 40 years a network of contacts have kindly assisted and supported my own research, writings and publications, and I am grateful for their unselfish help. It has also been rewarding to help each of them in their own research projects. Space sleuthing is a two-way process. What you might be looking for they might have and your information could be the key to their research. That is part of the fun of it all.

Notable in this select group have been: from the UK – Phil Clark, Rex Hall, Anders Hansson, Gordon Hooper, Neville Kidger, Andrew Salmon, George Spiteri, Rob Wood, Andy Wilson and Keith Wilson; from the Netherlands – Bert Vis, Anne van den Berg, Chris van den Berg and Gerard van den Haar; from Belgium – Bart Hendrickx; and from France – Daniel Tromeur; from Ireland – Brian Harvey; from the United States – Jim Oberg, Michael Cassutt, Nicholas Johnson and Bill Barry; from Russia – Vladim Molchanov; and from Australia – Colin Burgess.

We have lost the friendship, knowledge and skills of Rex Hall, Neville Kidger, Anne van den Berg, Chris van den Berg and Vladim Molchanov. All sadly missed but fondly remembered.

We lost Neville in December 2009. He had been suffering from leukemia for some time but his loss was sudden and heartfelt. I was told of this sad news by a phone call from Rex whilst I was out working. When I returned home I phoned my condolences to Neville's wife Wynn and his daughters. Without Neville's popularising, Russian space stations would not have generated the interest they did in countless readers of *Spaceflight*. He made each instalment interesting and easy to read, as well as accurate and informative. Without Neville, it is certain that *Zenit* would not have become the popular magazine it did. In May 2010 George Spiteri, who has followed the Soviet programme since the 1960s, took over Neville's role of Space Station correspondent for the BIS.

Rex Hall continued his work in education and youth development in London, and won a government contract in the late 1990s to use sport to motivate underachieving youths in skills such as literacy, numeracy and information technology. This 'Playing for Success' venture was the only UK-government maintained programme to achieve its promise and deliver what it set out to do. His sterling work earned Rex an MBE in 2003 for services to education. It was a credit to the man and the devoted support of his wife Lynn that this work remained a passionate priority totally separate from his intense interest in the Soviet-Russian space programme and dealing with his serious personal health issues. The last time I saw him was at the Autographica 15 show in early May 2010, and it was clear he was not well. Sadly Rex passed away at the end of May. He was only 63. He left a very large void in many lives but also filled those same lives with his enormous charm, wit, warmth and unique personality.

Rex had introduced me to the mysteries and sheer fun of sleuthing cosmonauts three decades before. Through Rex, I had the great fortune to meet many cosmonauts both in the UK and in Russia. I plan to produce a revised version of *Soyuz* in tribute to a true gentlemen, a great friend and a dedicated researcher.

The sleuthing continues

As a side issue to the sleuthing of the Soviet and Russian space programme there is a growing interest in the activities and scope of the Chinese manned space programme. With the similarity of Shenzhou to Soyuz, there is a viable reason to delve deeper to compare these two spacecraft. The secretive nature of the Chinese 'Taikonaut team' is similar to that of the early Soviet cosmonaut corps and for this reason has begun to attract the attention of several modern-day space sleuths, one of whom, Tony Quine of the Isle of Man, has performed detailed research into the selection of women into the Chinese team [7].

I unearthed an interesting fact on an earlier Chinese programme to enter space on the American Space Shuttle some years ago whilst researching the archive of former flight director Clifford Charlesworth at the Johnson Space Center. I found a couple of internal memos which indicated that a contingent of Chinese space officials would be touring the facility early in January 1986, and this group would include the "Chinese payload specialists". At that time there were rumours of an offer to fly a Chinese PS on a Shuttle mission, but the loss of Challenger later that month put an end to those plans. It has therefore been very difficult to determine the identity of this team, and who may have been on the short list to train for such a flight. The information is out there somewhere, so it is not a case of giving up but just keeping chiselling away at seams of information as they surface. Of course, the "payload specialists" referred to in the memo may have been members of the training team, touring JSC in order to gain experience in advance of assigning the real payload specialists. This was seen more recently when a pair of "Chinese training staff members" participated in cosmonaut training at TsPK in the late 1990s. They later turned up in the first group of Chinese Taikonauts, though they have yet to make a spaceflight.

Phil Clark has written several in-depth articles on the developing Chinese space programme over the years, publishing in *Spaceflight*, *JBIS*, *Zenit*, amongst others. Irish author Brian Harvey has written a definitive history of the Chinese programme which sets the mark for new volumes to document a fresh arena in space sleuthing – Chinese style [8].

CONCLUSIONS

There is still plenty to write about. I have been fortunate to turn a childhood hobby into a career, and as a result of considerable space sleuthing into both the Soviet and American programmes have had the pleasure (and luck) of seeing more than 20 titles published over a 25 year period – and there are more in the works. Dealing with both American and Soviet-Russian programmes sometimes blends into a single project. In addition, it has also been a pleasure to deliver a number of presentations, lectures and formal papers on my research into the Soviet and Russian space programme over the years, most of which, of course have been to the membership of the BIS but have also included other societies and social groups, professional and educational audiences.

It is clear that despite the growth of the Internet, online scanned documents, and global communications, nothing can beat plain old hard work in researching paper upon paper and document after document. Although often tedious and hard work, it pays off in the end and can be fun when a spark of new information comes to light.

There are undoubtedly more stories to be found and told for the space programmes of every nation, but those are for the future.

Above all, sleuthing the Soviets enabled firm friendships to be forged across the world, and that these have lasted decades has to be the greatest reward of all.

References

1. Kenneth Gatland, *Manned Spacecraft*, Blandford Press 1967.
2. David Shayler, 'Call-signs of Soviet Manned Spacecraft', *Spaceflight* January 1977 p.6.
3. 'Sir Patrick Moore medal for Neville', Society News, *Spaceflight* February 2004 p.87.
4. David Shayler, 'Where Blue Skies turn Black: Soviet Stratospheric Balloon Programme of the 1930s', *JBIS* Vol.50 No.1 p.8 (1997).
5. *From the Flight Deck 2: NASA Space Shuttle* with Harry Siepmann (Ian Allan 1987). *Apollo 11 Moon landing* (Ian Allan 1989). *Challenger Aviation Fact File* (Salamander Books 1987).
6. Rex Hall & David J. Shayler, *The Rocket Men: Vostok and Voskhod, the First Soviet Manned Spaceflights* Springer 2001. *Soyuz: A Universal Spacecraft* Springer 2003. Hall, Shayler & Vis, *Russia's Cosmonauts: Inside the Yuri Gagarin Training Center* Springer-Praxis 2005.
7. Tony Quine, 'Identity of female taikonaut trainee confirmed', *Spaceflight* January 2011 p.5.
8. Brian Harvey, *Chinese Space Programme*, John Wiley & Sons (1998), *China's Space Program* Springer (2004), *China in Space* Springer-Praxis (2013).

Contributors

Phillip Clark was born in Bradford, Yorkshire, in 1950 and became interested in spaceflight after listening to the early NASA manned missions at school. He focused on Soviet spaceflight at the time of the Apollo 11 mission and began corresponding with Geoffrey Perry of the Kettering Group. Clark has an Open University (OU) degree in mathematics and computing, and pioneered the use of computer analysis to uncover the roles of obscure Soviet and Western reconnaissance satellites from their orbits. For many years he was a space consultant for the BBC and is the author of the 1988 book *The Soviet Manned Space Programme*. He currently lives in Hastings, England.

Sven Grahn is from Stockholm, Sweden, and as a teenager helped launch sounding rockets from the Kronogård rocket base. He holds a master's in engineering physics and joined the Swedish Space Corporation in 1975. Grahn was the project manager for Sweden's first microgravity project (a module for the German TEXUS rocket) first launched in 1977, deputy project manager for Sweden's first satellite VIKING (launched in 1986), and engineering manager for the first satellite entirely designed and integrated in Sweden – the FREJA magnetospheric satellite launched by China in October 1992. From 1993 until 2001 he headed a Swedish Space Corporation team designing sounding rocket payloads, balloon gondolas and small satellites. Between 2001-2006 he was Senior Vice-President of Engineering at the corporation.

Brian Harvey is a writer, broadcaster and journalist based in Dublin, Ireland. He has a degree in history and political science from Dublin University (Trinity College) and a master's in economic and social history from University College Dublin. His first space book, *Race into Space* (Ellis Horwood, 1988), was a history of the Soviet space programme. He has since written histories of a number of the world's space powers, paying close attention to China. His *Russian Space Probes* (co-authored with Olga Zakutnyaya, Springer-Praxis, 2011) is a history of Russian and Soviet space science.

Bart Hendrickx was born in Kapellen, Belgium, in 1964. An early fascination with Russian spaceflight developed into an interest in the language. He has a master's in Dutch-English-Russian translation in 1986. Hendrickx is a full-time language teacher at the University of Ghent. He has written extensively on the history of the Soviet space programme, primarily based on Russian-language sources. He is the co-author (with Bert Vis; Springer-Praxis, 2007) of a history of the Soviet space shuttle.

Christian Lardier was born in France in 1952 and joined the Cosmos Club de France aged nineteen. Now a professional journalist, he has been the space editor of *Air & Cosmos* magazine since 1994. He is also a senior member of the Association Aéronautique et Astronautique de France, co-founder of Association Planet Mars, co-founder and president since 2007 of the Institut Français d'Histoire de l'Espace, and a member of the history committee of International Academy of Astronautics.

James Oberg was born in New York in 1944, and in 1969 got a master's in applied mathematics from Northwestern University. After service with the US Air Force he joined NASA in 1975 and worked at the Johnson Space Center until 1997. As a child 'space nut', Sputnik inspired his early private study of Soviet spaceflight, an interest later encouraged by mentors such as Charles Sheldon of the Congressional Research Service. He became one of the original sleuths in his spare time after writing dozens of articles, authoring the ground-breaking book *Red Star in Orbit* and circulating a sleuthing newsletter called *Cosmogram*. At JSC he was an orbital rendezvous mission controller, receiving professional awards for his work relating to the ISS orbit. He has authored ten books and is currently a consultant for NBC News, which recently took him to North Korea's secret launch site. He lives in rural Texas with his wife Alcestis ('Cooky') Oberg, herself a published author.

Dominic Phelan was born in Dublin, Ireland, in 1972. After gaining a qualification as a journalist in 1996 he worked as a freelance writer with features published in *The Irish Times*, *The Irish Independent* and *History Ireland*. His articles on the history of astronomy and spaceflight have appeared in *Spaceflight* and *Astronomy Now*, and he contributed an 11,000-word chapter on Soviet lunar plans during the Moon Race for *Footprints in the Dust* (University of Nebraska Press, 2010). He has travelled to Moscow and written about Russian medical preparations for a manned Mars mission that was used in *Space Exploration 2008* (Springer-Praxis, 2008). He has attended the annual BIS Soviet Forum since 1993.

David Shayler was born in Birmingham, England, in 1955. His interest in space was kindled while at school. He trained as an engineering draughtsman, then joined the Royal Marines. On returning to civilian life in the late 1970s he worked in various retail management roles before going full-time as a space promoter with his company Astro Information Service in 1999. As part of a drive to promote 'space education' amongst the public, he has given several hundred lectures and is the author of over twenty spaceflight books. He is a Fellow of the British Interplanetary Society and co-founder of the Midland Spaceflight Society. He lives in the West Midlands with his wife Bel.

Asif Siddiqi is the author of *Challenge to Apollo: The Soviet Union and the Space Race, 1945-1974* (NASA History Office, 2000), the first comprehensive work on the history of Soviet spaceflight to be published after the opening of the former Soviet archives. He was the series editor of Boris Chertok's four volume *Rockets and People* memoir published in English between 2004 and 2012 by NASA, and he authored *The Red Rockets' Glare: Spaceflight and the Soviet Imagination, 1857-1957* (Cambridge University Press, 2010). He is currently Associate Professor of History at Fordham University in New York. In 2013-2014 he will serve as the Charles A. Lindbergh Chair in Aerospace History at the Smithsonian's National Air & Space Museum in Washington, DC.

Bert Vis is from Voorburg in The Netherlands and was born in 1955. His interest in manned spaceflight began with the launch of Apollo 7, and by the mid-1970s he was corresponding with NASA astronauts and Soviet cosmonauts seeking information on their careers. Since the 1980s he has visited the USA and Russia on an almost annual basis and is also a regular at the Association of Space Explorers congress. In addition to articles for Dutch and international space publications, he has contributed to two editions of *Who's Who in Space*, written a chapter for *Fallen Astronauts* (University of Nebraska Press, 2003), and co-authored *Russia's Cosmonauts* (Springer-Praxis, 2005) and *Energiya-Buran* (Springer-Praxis, 2007). Vis served as a firefighter in The Hague for thirty-three years and is now employed by the Haaglanden Regional Fire Department as a policy adviser.

Claude Wachtel was born in Paris in 1951. He studied space geophysics at Pierre and Marie Curie University in Paris and joined the staff of the Meudon Observatory, where he works on the results of experiments carried on satellites. He has a doctorate in space geophysics and has worked in the Geodynamics and Astronomy Study and Research Centre in Grasse, South of France. In 1978 he went to work in the French Prime Minister's office on risk contingency planning and crisis management. He has organised expeditions to the Arctic, holds an aerobatics qualification, and gained the *Légion d'honneur* in 1987.

Index